FRANK BUNKER

CW00338312

BRICKLAYING
SYSTEM

Elibron Classics
www.elibron.com

Elibron Classics series.

© 2005 Adamant Media Corporation.

ISBN 1-4021-6113-1 (paperback)
ISBN 1-4021-5354-6 (hardcover)

This Elibron Classics Replica Edition is an unabridged facsimile
of the edition published in 1909 by the Myron C. Clark Publishing Co.,
New York and Chicago.

Elibron and Elibron Classics are trademarks of
Adamant Media Corporation. All rights reserved.

BRICKLAYING
SYSTEM

BY

FRANK B. GILBRETH
M. Am. Soc. M. E.

NEW YORK AND CHICAGO
THE MYRON C. CLARK PUBLISHING CO.

LONDON
E. & F. N. SPON, LTD., 57 Haymarket
1909

Press of
THE WESTERN NEWSPAPER UNION
CHICAGO

TABLE OF CONTENTS.

LIST OF ILLUSTRATIONS.

INTRODUCTION.

The art of bricklaying is unique in that the fundamental principles of brick work today are not unlike those exemplified by the oldest ruins of thousands of years ago. The bricklayer also stands almost alone, as one who has not been obliged to compete in his trade with women, with machinery, or with foreign manufacturers.

Each old country has had its local methods of bricklaying; but not until the last score of years have bricklayers, coming to America from all countries, bringing their best local methods with them, learned from each other and adopted the best of each others' methods.

The purposes of this book are as follows:

(a) To put in writing that knowledge which has been handed down by word of mouth from journeyman to apprentice for generations.

(b) To record methods of handling labor, materials and plant on brick work that will reduce costs and at the same time enable the first-class workman to receive higher pay.

(c) To enable an apprentice to work intelligently from his first day, and to become a proficient workman in the shortest possible time.

CHAPTER I.
TRAINING APPRENTICES.

1

There is no immediate profit from apprentices unless large numbers of them are employed on the same job at the same time. As large numbers of apprentices invariably cause trouble between the employers and the bricklayers, it is necessary to limit our apprentices to those boys who, when they have become trained, will make valuable additions to our organization.

2

Hire only those apprentices who will apparently make good foremen, unless bricklayers are scarce.

3

Two or more apprentices on the same job work out better than one, as there is a spirit of rivalry between them, and they can be matched against each other in speed contests.

4

The term of apprenticeship shall be at least three years; additional time for lost time and vacations.

5

Apprentice shall not be permitted to work without overalls until he is out of his time.

6

The first day that an apprentice is put to work he is to be provided with a brick hammer and trowel at our expense. Procure an old trowel that has been broken in by some good bricklayer on the job. The foreman bricklayer should make it his special duty to see that the trowel is slightly undersized, also the best and the handiest trowel on the job. He should give the bricklayer a new trowel.

7

At the end of six weeks, if the apprentice has done well, he is to be given a new spirit plumb rule with two plumb glasses and one level glass. This plumb rule should be 3 ft. 6 ins. long.

8

As soon as he has progressed far enough to warrant it, he shall be given another large trowel, brick set and jointer. He shall furnish himself with everything else that he needs.

9

An apprentice should be taken in charge by an intelligent bricklayer, who should be responsible for his actions and work, for a period of one week. He should be put under a different bricklayer every week for at least a month. At the end of that time he should be put on that part of the work where he can earn his money, and at the same time learn the most. In other words, we do not want our apprentices to be kept on heavy work, if they are constantly doing their best. At the same time, we do not want them promoted any faster than they can earn their money.

10

Apprentices must be worked to their full limit of endurance.

11

Apprentices must not be hazed nor misled after their first day at the trade. Foremen must answer every question that they ask in good faith, regardless of how simple it may seem.

12

An apprentice is supposed to do a man's amount of work on filling in the middle of the wall after the first month. He is supposed to do a man's amount of work on all common brick work after six months.

13

First of all, an apprentice should be taught that all brick, even common brick, have a top and a bottom, an inside and an outside.

14

The outside is generally determined by the way the bricks have been stacked in the kiln, but if the natural outside is chipped, oftentimes the natural inside is the better side.

15

All bricks made by hand in a mold are a little wider at the top or open side of the mold than at the bottom of the mold. This is sometimes caused by the molds being made slightly smaller at the bottom than at the open top, sometimes by the contact of the front and back side of the brick being slightly distorted by contact with the sides of the mold as the soft wet clay slides out of the mold.

16

In hand made brick, and in nearly all brick except wire-cut brick, the natural top can be told from the natural bottom because it is much rougher. The top can generally be easily told from the bottom by feeling, if the brick is held in the usual position used just before laying. The bottom being narrower than the top, the brick can be held by less pressure of the fingers on the front and back of the brick when it is right side up than when it is upside down. A few minutes will enable a novice to detect a very small difference in the width of the top and bottom of a brick, by observing how much pressure of the fingers is needed to sustain the brick.

17

Brick must be laid with the wide surface uppermost, like Fig. 1 (exaggerated). If the front and rear faces of the brick are parallel, such as is the case with wire cut brick, then the brick must be laid like Fig. 2 (exaggerated). All bricks must

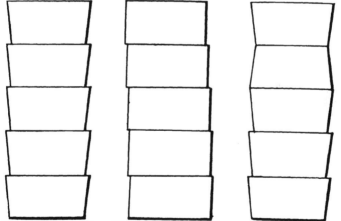

Fig. 1.—Hand Made Brick.   Fig. 2.—Wire Cut Brick.   Fig. 3.—One Brick Upside Down.

be laid with slightly overhanging faces, so that the appearance of the wall (exaggerated), will be similar to beveled siding or clapboards upside down. The reason for this is that the bricks vary greatly in thickness, and one edge only is laid to line. This is the top edge. The lower edge is not laid to line. It is, therefore, not so straight, and is set slightly inside the line of the top edge of the course below it to hide its inaccuracies. While the amount of overhang and set-in of each course is slight, it is enough to show very plainly. If you sight down the face of the wall you will see nothing but mortar. If you sight up the wall you will see nothing but brick. If a course is laid with considerable overhang it is called "rolled." If a course is laid with a batter it is called "lipped." See Fig. 3.

18

One of the worst mistakes that can be made in the training of an apprentice is to expect him to do perfect work first, and fast work later. A boy taught after this scheme is sure to get into bad habits of laying brick with too many unnecessary motions that will prevent him from ever laying brick fast. This is very important.

19

The right way is to put the apprentice at work where the appearance of the work is not of importance. Insist that he lay as many brick as a journeyman, even if they are not laid quite so well. Teach him to lay a brick with the least possible number of motions, and, instead of correcting all of the little faults on one brick, to try to lay the next brick without the same faults as attended the laying of the preceding brick. This last method will teach speed, and skill will surely soon follow, with sufficient practice.

20

Of course it is not intended by these directions that an apprentice be permitted to do any work that would affect the stability of the work. It is simply a matter of looks, and he must start where looks are not important.

21

"Motion study" is of the greatest importance in teaching a trade quickly. It is also the most profitable method of teaching a trade.

22

An apprentice must be made to lay brick with quick motions, even on his first day. Speed and the least number of motions must be uppermost in his mind at all times. The apprentice must be made to lay brick with the method outlined in this system even if it is necessary to have a bricklayer go over his work as fast as he lays the brick, to make his work right.

23

As soon as an apprentice has formed a fixed habit of laying brick with only a few, and with no unnecessary motions, he must be constantly reminded that the quality of the workmanship will be remembered long after all other considerations are forgotten.

24

The rules in this system must be construed as solely for the purpose of eliminating unnecessary delays, unnecessary labor, and unnecessary expense, and never for any short cuts that produce speed, economy or profit at the expense of the best workmanship.

25

Leave the so-called "tricks of the trade" to those persons

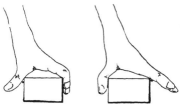

Fig. 4.—Laying Brick Without Disturbing the Line.

who have not served a proper apprenticeship and who do not know how to do the best work.

26

Show the apprentice how to lay a brick without disturbing the line. Impress upon his mind that his fingers must not even touch the line, or the line will be pushed out of place while the other men are trying to use it as a guide. See Fig. 4. Make him practice before laying to the line, so that his thumb and fingers will come up as the brick goes down near the line. Do not let him lay brick on the line until he can do this without disturbing the line.

27

An apprentice should be taught to hold his trowel like a razor, with two fingers only, and with the thumb on the top of the handle. The thumb should never be put around the handle of the trowel. The handle of the trowel should be kept perfectly clean at all times.

28

Regardless of the locality in which an apprentice works, he should be taught to lay brick both "Eastern" and "Western" methods, not only so that he can have the experience for his own use, but also that he can boss both kinds of bricklayers to the best advantage.

29

The "Eastern" method is to "pick-and-dip" brick and mortar at the same time; the mortar being in a tub, or in a mortar box with beveled sides.

30

The "Western" or "stringing mortar" method is to use a much larger trowel than could possibly be used in a tub. This necessitates a mortar board or mortar box, and first spreading mortar enough for several brick ahead, and then picking up only brick.

31

Each method has its advantages and disadvantages. Some conditions make the "Eastern" method preferable, some the "Western" method.

32

If the apprentice is taught both methods, he will know, instinctively, which is the better to use under varying conditions. The kind of sand, the proportions of the cement, lime and sand, the dryness of the brick, the methods employed by the men on the leads—all these go to determine which method will give the most speed, economy and quality.

33

When an apprentice reaches to pick up bricks, see that he picks them up with both hands at precisely the same time.

34

When he reaches for mortar with one hand and brick with the other, teach him to pick up both at the same time. He should look at the mortar as he starts to reach for it, but he should pick it up by the feeling, and his eyes should be only on the brick that he is picking up with the other hand at the same time. See Fig. 5.

35

To appreciate fully the importance of this rule, watch several bricklayers a few moments. You will notice that the man who unconsciously is picking up with both hands at the same time, can do his work faster, and with much less effort to himself, than he who is picking up first with one hand and then with the other. This is largely a matter of habit. If an apprentice is allowed to pick up first with one hand and then with the other, it will be hard to break him of it.

36

The bricks nearest the wall should generally be picked up first, so as to maintain a clear place to stand and a clear

Fig. 5.—Picking Up Stock with Both Hands at the Same Time.

place for the tenders to walk, in case they find it necessary to pass on the scaffold.

37

An apprentice should, at first, be taught to throw only enough mortar for one brick at a time on the line. After he is fairly proficient, he should be taught to throw his mortar for at least two brick, even when laying the pick and dip method. See Figs. 6 and 7. He can pick up mortar for two or three bricks just as quickly as he can pick up mortar for one brick. With practice, he can throw the mortar for three brick just as quickly, as he can for one brick, and he can certainly lay the extra one or extra two brick faster if he does not have to dip and throw the mortar more than once to two or three brick.

38

An apprentice should be taught that the bricklayer depends on sighting with his eye to get plumb corners nearly plumb, and that the plumb rule is used to correct the inaccuracies of his eye. The first three courses should be plumbed accurately to furnish a guide to the eye. A corner can be "sighted" quickest and most accurately, by sighting one side at a time, i. e., putting the eye exactly over the corner to be plumbed and moving the eye in the plane of one face of the corner, and in the direction away from the wall. Moving the eye back and forth to a point a slight distance from and exactly plumb with the corner, will make the corner apparently

Fig. 6.   Figure 7.
Throwing Mortar for Two Brick.

change from a point to a line. Any inaccuracy from a straight line is quickly detected. After one face of the corner has been corrected, sight the other face in the same manner. When the two faces are plumb, the corner will be plumb. Do not try to make the corner appear straight both ways at once. The method is slower, and it may appear straight when in reality it is not.

39

It is a simple matter to describe to the apprentice the best methods of laying brick for the greatest strength. It is a very difficult matter to explain to him how to lay pressed face brick in a manner that will make a large, plain, blank wall appear accurate and uniform under the most critical examination.

40

The following rules will help the apprentice to make the best appearing work on pressed brick face work:

(a) Use the thinnest line obtainable, that will stand a hard hauling without sagging or breaking.

(b) Build small leads, so that as much of the wall as possible is built to line, instead of built as a lead.

(c) Make the line fast around the end of the wall, and wound around a brick on the lead, so that a tight hauling will not pull down the lead. Another reason for doing this is that if the nail is used in the lead and the lead pulls down, some bricklayer may lose an eye by the nail flying through the air.

(d) See that the line is placed 1-32 in. outside the top edge of the brick and exactly level with it.

(e) See that no brick touches the line.

(f) See that the line is disturbed as little as possible when laying a brick.

(g) Do not lay a brick that is thinner, thicker, shorter or longer than the others, even if it is of the same cull. Use it for filling.

(h) Use the right amount of mortar. If you use more than the right amount it will squeeze out, and daub the brick underneath it.

(i) See that the lower edge of the brick is distinctly back of the line of the top edge of the course under it. The amount that it should be back varies with the brick and the conditions. It should never be less than 1-32 in. and seldom more than ⅛ in. Work looks decidedly better with too much set in (or roll) than not enough. This is one of the most important rules for good looking brick work.

41

The work of the apprentice differs from that of the journeyman in appearance largely because the latter is able to make the set in, or roll, or overhang, of all brick exactly alike.

42

An unskilled man can use a story pole to get brick courses the right height. He can use a plumb-bond pole, and mark exactly where the end joints should come. He can use a tight line, putting the top edge of each brick to it. He can cover up inaccuracies with good jointing and hide the differences of the thickness of the brick by skillful ruling. He can make every brick touch a plumb rule and straightedge, by tapping back the projecting brick with a hammer before the mortar has set. Yet the wall will have a bad appearance unless it looks uniform and each brick has the same amount of set-in. There is no way of correcting the amount of roll after the bricks are laid.

43

This is where skilled practice counts. It makes for uniformity. This is where the journeyman makes the good looking wall and where the apprentice finds difficulty and must lay the most stress; for when the sun moves from a position in the plane of the face of the wall and begins to throw long shadows on that wall, the inaccuracies are greatly exaggerated by the sunlight and the shading, due to lack of uniformity of set-in and roll.

44

Apprentices must study carefully the rules, photographs and charts in this system, especially those relating to methods. They must observe the work done by the various bricklayers, and must, after study of this system, the work they see, and their own work, make out charts of their own processes.

45

They must comply with all rules of the Field System, as well as of this Bricklaying System, and must be made to realize that they are a part of the organization, and that a knowledge of it and the systems by which it is operated, are absolutely essential to any advancement.

46

Apprentices must be shown that the bricklayer's trade is one of the oldest, most respectable and most desirable of trades, and one worthy of the entire attention of any bright, educated and determined American boy; that the knowledge gained of the trade can never be taken away from the one who has once obtained it, regardless of what ill fortune has overtaken him; and that $25 to $100 per week will always stand ready for the man who can lay brick, or who has sufficient knowledge of the trade to supervise the work of other bricklayers.

CHAPTER II.
METHODS OF MANAGEMENT.

47

Foremen will be rated and paid according to the quality and not according to the quantity of work that they secure from their men. Not only on account of the greater pleasure that all derive from doing the best work, but also as a business proposition, the most permanent success will come from having earned and deserved the reputation of doing the best work.

48

The foreman must see that all work is laid out in a systematic manner. The men should be so selected and grouped that, the job once started, speed and efficiency will be apparent and can be fittingly recognized.

49

On small jobs a foreman may know exactly what each bricklayer is doing every minute during the day. On large jobs, if the foreman knows this, it is because he is neglecting to look after some points that are of much more importance.

50

A foreman should study to arrange his men so that the work of the slow men will show up automatically to their disadvantage. This can be arranged in many ways, one of which is to divide the gangs into units, the number of men in a unit to be determined by the character of the work.

51

Take, for example, a wall of nine piers, separated by eight windows. On this wall there should be nine bricklayers, if the piers are of about the same size. If the piers are not the same size, the number of bricklayers should be increased or reduced, so that their work will be equal, and the slow man will be shown up quickly. The foreman should watch the bricklayers to see which man is standing up idle. He is standing up for one of three reasons: (a) he is loafing; (b) he is out of stock; (c) he has finished his bit.

52

If it is for the first reason, he should be dealt with. If he is out of stock, the leader of tenders needs attention. If it is for the third reason, his speed should be recognized, and the mason who is behind and delaying the raising of the line to the next course should be investigated.

53

It is sometimes difficult to divide a wall that is not symmetrical into equal parts so that each bricklayer will have the same sized section; but with a little study almost any wall can be divided fairly. For example, if the wall is cut up by openings so that the piers are not the same size, it is often economical to put fewer men on the wall and apportion several piers to each man. In this way it can be divided evenly enough to make conditions favorable for a contest. Sometimes the trig can be put enough off center of the wall to compensate for some extra plumb work in a break or chimney flue. Again, a small pier in the wall may be assigned to an apprentice.

54

On a long wall, it is often economical to provide one or two special men to take care of a large irregularity in the wall, and to hold the contest on the remaining straight or symmetrical parts of the wall.

55

In any case, the foreman should watch the first few courses as laid, then shift the men enough to make the stints as nearly equal as possible.

56

The work of a bricklayer is generally indicative of his personal character. If he is dishonest he will do dishonest work and cover it up if possible before the flaw is seen. If he is honest he will leave his job before he will do scamp work, even at the suggestion of his foreman. If a bricklayer is ever caught doing a scamp piece of work he should never be absolutely trusted again.

57

Sometimes the athletic contest spirit and desire to outclass the others leads some of the bricklayers to do careless work, especially where it cannot be seen, as in the middle of the wall. One of the best methods of counteracting this is to write the name of each bricklayer on the plan, showing where each worked, and to let the bricklayers see that their names are being written on the plans. There is, of course, nothing new about this scheme, as stone masons for centuries have put their own marks on stones. The name or mark on the work undoubtedly makes the workman take more interest in his work.

58

On engine beds and similar work, where the pieces are isolated, assigning gangs of men of different nationalities to the different beds will create extra interest in the contests. If this is not feasible, put the tall men on one bed and the short men on the other, or the single men against the married men, or the eastern "pick and dip" men against the western "string mortar" men.

59

While one who is not experienced at making his men really enthusiastic on their work cannot appreciate how athletic contests will interest the men, it is the real secret of the success of our best superintendents. It not only reduces costs, but it makes for organization, and thus saves foremen's time.

60

There is no way that continued interest in athletic contests can be maintained so well as by having a fair and correct score kept of the results of the labors of the different men.

61

When it is not possible to divide the work so that each man's work shows up all by itself, the best arrangement is to divide the men up into two or more gangs of as few men as possible, generally with two, four, six or eight men to a gang.

62

If the character of the walls is similar, but if they are not the same dimensions, it is wise to provide some other form of measurement than the height of the wall, such, for instance, as a score on a large black board, so placed that it can be seen by all the men of all the gangs. This board should be ruled off, and the score should be carefully filled out.

63

The men can see the score and the contest can be carried out throughout the entire day. In the case of extra pay being

given for particularly high records of brick laid, the bricklayers can see every half hour just how successful their ef-
forts are.

64

In order that the bricklayers may always have a square
deal, the method of estimating the number of brick that they
lay per day shall be as follows:

65

The story pole on the hauling end of the line shall be
marked off in courses, as usual. At each mark, the number
of brick in that course shall be plainly marked on the pole. In
case of any dispute, any one can then check up the records,
by actually counting the number of brick in each course.

66

As a general rule, the men should be separated so that
the amount of their individual work will show up separately.
This will bring about the best results, whether or not there is
a well organized athletic contest in progress.

67

It is seldom good practice to have the tenders work indi-
vidually instead of in small gangs, due to the difficulties of
passing on narrow runs and foot stages.

68

Do not permit your hod carriers, wheelbarrow men or
packet men to come up one at a time.

69

Have those carrying any one kind of material that are
tending masons on any one wall fill their hods all at once,
shank their hods all at once, and start all at once, but do not
have them drop their hods at exactly the same instant, as
it might endanger the stability of the scaffold.

70

Hold the leader responsible for the work of the entire
gang. They must dump their material where the leader says
they must dump it. They must never throw the brick from
the hods in a manner that will scatter or break, or even chip
the brick. They must empty the hod in a manner that will
enable the bricklayer to pick up the brick the easiest, as bricklayer's time costs so much more than hod carrier's time that
the hod carrier can afford to waste two minutes of his time
any time that it will save one minute of the bricklayer's.

71

It may seem at first that making the men all fill at once and all start at once, and all go at once, is like holding the entire gang back to the speed of the slowest man. This is not so. On the contrary, it shows up the slowest man and he can be removed.

72

Have the men keep their place in the line. Have the last man who is set to work with the gang be the next to the last man in the line.

73

Pick out a good man for leader and pay him 10 per cent more than the rest, because he is expected to direct as well as work. He must see that he has more men when the bricklayers are backing up than when they are laying overhand.

74

In the event of the usual leader being absent, promote the rear leader to leader with leader's pay and the second man in the line to rear leader with 5 per cent more pay than the rest. This will make all the men in the line desirous of the leader's job with leader's pay. It will, if the rules are carried out, make the organization of the tenders automatic. The foremen will only have to watch the leaders in order to handle all the tenders.

75

The same system applies to wheelbarrow men.

76

Do not put brick tenders and mortar tenders under the same leaders.

77

Piling brick into a wheelbarrow is a matter that requires considerable attention. It is not enough to tell the tenders to fill the barrows with brick. The barrow gang must be shown how to do it in the quickest and least fatiguing manner.

78

The leader shall be paid at least 10 per cent. more than each of the rest of the gang. He shall be carefully shown that the barrow must be placed as near the brick as possible. He must not be allowed to be leader unless he continually picks up brick with both hands at the same time and fills his barrow faster than the other men. He must convey the brick from the pile to the wheelbarrow in the shortest possible line, and both brick must be put in the barrow at the same instant.

79

It will be difficult to get every barrow man to abide by these rules, but the leader must be made to follow this method, and the extra money paid the leaders is an incentive to the others to work to be promoted to be leaders of gangs.

80

The men must be taught to pile the brick so that the load comes over the wheel, instead of on the legs of the wheelbarrow. This enables the man to wheel larger loads.

81

It should be decided by the brick foremen how many brick shall be put in the barrow by the leader, and he must count his load every time. The leader's load should be checked up occasionally, especially when his load appears to be small and if it is found short, he should be replaced by another leader. The number of brick, of course, depends on their dry weight, with due allowance when they are wet, and on whether the bricks are to be wheeled up an incline or on a level.

82

Rear end leaders are necessary on long trips where the last man leads the way back. Where the lap is continuous, a rear end leader is not necessary.

83

A rear end leader should receive 5 per cent. more than regular tender's pay. He should be selected from the line of tenders, with the idea of promoting him to leader if he handles the gang promptly.

84

A properly organized and trained gang of tenders will do from 50 per cent. to 200 per cent. more work than an untrained gang.

85

When unloading cars of materials, try to have but one man to a car, and start several men at the same time. If this is not possible, start two men at the same time, but at opposite ends of the same car. This will enable the foremen to pick out the first-class men.

86

Reward the winner every time.

87

Provide shields of No. 10 gage sheet steel for men to shovel on especially when unloading cars, or when handling a large amount of dumped material, such as sand, coal, etc.

88

The shields are to be of the size and shape shown by Fig. 8, and are to be used whenever possible as a flat surface on which to shovel.

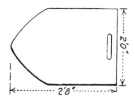

Fig. 8.—Shield on Which to Shovel.

89

The hole in the shield should be made large enough to permit the handle of the shovel to pass through it. The shield can then be easily carried from car to car on a man's back, suspended from the handle of the shovel which he carries over his shoulder.

90

Too much stress cannot be laid on the necessity and value of having only first class men. Athletic contests have proved the surprising fact that first class men ordinarily do twice to three times the amount of work of other men whose methods of working disguise their slow pace until the athletic contest shows them up.

91

The difference in cost between the best work and the worst work is such a very small amount as compared with the entire cost of the undertaking that no firm or individual can afford to be identified with any but the best class. The features that increase the cost of work materially are, not working the men to advantage, and having them remain idle for any reason, such as for want of stock, lack of incentive for large output, lack of proper superintendence, etc.

92

Quality of the work must be given preference over quantity of output at all times.

93

The winners of the athletic contest should be paid higher wages than the rest of the men. They should furthermore be given first opportunity to make overtime wages and they should be kept till the last to finish the job.

(For further suggestions regarding athletic contests, see "Field System.")

CHAPTER III.
METHODS OF CONSTRUCTION.

94

The planning of the methods of construction of a building should be laid out as carefully as the building of a great machine in a modern machine shop. Each part should be routed to its final place with the least handling and confusion possible.

95

The routing and the consecutive order in which each wall and each structural member of the building is to be built must be diagrammed, and the dates on which the materials are to arrive on the site, and to be put in place must be agreed upon by the purchasing department and the superintendent.

96

The method of attack on jobs where great speed is required must necessarily differ greatly under different conditions.

97

Figures 9 to 14 show progress of work on the Augustus Lowell Laboratory of Electrical Engineering for the Massachusetts Institute of Technology, which we built in Boston in two months and seventeen days during the summer vacation of 1902. This building, covering over 44,000 sq. ft. of land was completed 54 days ahead of contract time. It is a particularly good example for study. It shows that speed can be obtained by organization, system, and the elimination of unnecessary delays, without slighting the workmanship, by simply planning ahead the date of arrival of materials and their proper routing.

98

The contract for this building was signed June 28th, 1902. The next two days were spent in studying the method of attack. Various schemes for dividing the building into small units were considered. The one finally adopted was to di-

19

vide the building by horizontal planes into units, with a foreman in charge of each unit. The various units were:

1. Laying out work, staking piles, giving levels, etc.
2. Piling.
3. Excavation, bracing of trenches and sawing off piles, pumping, rigging.
4. Concrete around piling.
5. Block granite foundations.
6. Brick work up to first floor.
7. Structural steel.
8. First floor.
9. Brickwork first floor to roof.
10. Woodwork first floor to roof.
11. Woodwork above roof.
12. Sheet metal work.
13. Roofing.
14. Interior lath and plaster.
15. Heating and ventilating.
16. Plumbing.
17. Wiring.
18. Interior finish.
19. Painting and glazing.
20. Flooring.
21. Blackboards.
22. Furniture.

99

The success of this method was largely secured by starting
the different units at one corner of the work and running
out in two directions at once, as the line of front on each unit
gets longer until the unit is over half done, then it gets
smaller gradually.

100

Each foreman kept his workmen as long as he needed
them, then turned them over to the foreman of the other units
above. In other words, the duty of a foreman of a unit was to
complete his particular unit as fast as the unit under him
would permit.

101

The lathing was put on while the roof boarding was being laid, and the plastering proceeded simultaneously with the
progress of the roofing.

102

It is obvious that to get the most speed on a building,
as many different branches of trades must be put to work
on the building, and at as early a date, as is possible.

103

For a building of this character, the most economy can be
had by starting on a corner and enlarging in both directions as
fast as men can be set to work.

104

Figure 9 shows the work two weeks after beginning construction.

105

Figure 10 was taken three weeks after commencing the
contract. Two pile drivers are on the job, and the foremen
of units 2 to 8 are pressing each other right up to the
pile drivers. Foreman of unit 9 is starting. Note that
the scaffold is being erected on one side of the wall, while

the bricklayers are building the wall staging high from the opposite side. This method enables the bricklayer to shift over to the scaffold on the other side of the wall, and to continue immediately on his own part of the wall.

106

Never shift bricklayers around any more than is necessary. They are not proud of, nor interested in, a piece of work, unless they build it all. Furthermore, there is always a

Fig. 9.—Massachusetts Institute of Technology. July 13, 1902.

question as to who did the bad piece of wall if several different men worked on the same piece.

107

Figure 11 shows the building four weeks after commencing work. Note that the roof is being put on in one corner, while the excavation is under way in the diagonal corner. The laths are being put on and the plastering mortar is being made up ahead.

108

Figure 12 shows the building five weeks after commencing work. From the beginning of the work 300,000 bricks, ahead of what were needed, were always kept on hand, in or-

Fig. 10.—Massachusetts Institute of Technology. July 20, 1902.

Fig. 11.—Massachusetts Institute of Technology. July 27, 1902.

Fig. 12.—Massachusetts Institute of Technology. Aug. 3. 1902.

Fig. 13.—Massachusetts Institute of Technology. Aug. 10, 1902.

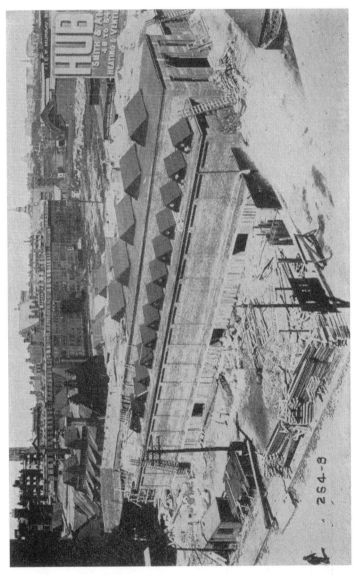

Fig. 14.—Massachusetts Institute of Technology. Aug. 17. 1902.

der not to run short, regardless of what might happen on the railroad from the brick yard. The best looking loads were hauled to the reserve pile. From this pile of brick the outside brick were culled. The rest of the brick were delivered as near as possible to the walls where they were to be used. Note the athletic contest on the four similar walls.

109

Figure 13 shows the building six weeks after commencing work. The roofing is started and the plastering is being put on inside the building. Note the athletic contest on the skylights, also the large number of ladders. Plenty of ladders and stairs will save traveling time on a building.

110

Figure 14 shows the building seven weeks after commencing work. Three-quarters of the roof is tight. Note the small amount of room that the Gilbreth scaffold requires when it is not in use. The 36 horses in the foreground will stage 350 lin. ft. of wall 22 ft. high.

111

These pictures prove conclusively the value of the method here employed.

CHAPTER IV.

ROUTING OF MATERIALS.

112

In order to secure the most effective work it is necessary that the material be routed to the men with the greatest economy of time and labor.

113

The transportation devices and apparatus to be used should be the subject of study while the plans are being drawn, so that after work on the job has once commenced no one need ever be held back by lack of an ample supply of conveniently placed material.

114

The foreman must secure from the office the plant lay--out planned for his job.

115

On the job shown by Figs. 15 to 17, the main building of which covers eight acres and which is a good example of a large undertaking, a great deal of study was given to the matter of routing the materials from the railroad spur tracks to the workmen.

116

After many kinds of mechanical methods were considered, we finally decided to use horses and carts, to load the carts at the cars on the spur tracks, and to haul the bricks to the exact place where the bricklayers could pick them up.

117

In order to accomplish this, inclined runways, Fig. 16, were built on various levels to the top floor of the building. The bricks were, by this method, carried from the cars on the track to the bricklayers with but one handling, that handling being the unloading of the cars.

118

Laborers were used to keep the brick in close to the masons, and to carry brick up to the scaffolds. This method reduces the number of tenders to a minimum. The mortar was

Fig. 15.—Methods of Routing on an Eight-Acre Job.

Fig. 16.—Specimen Driveway to all Floors.

transported the same way. The same runs were also used by the carts that hauled up the timber and flooring.

119

The organization on this work consisted of one general superintendent, and a foreman mason and foreman carpenter on each building.

120

A tower, Fig. 17, was built in the center of the lot and from this tower the superintendent could see all portions of the work. It was equipped with a telephone, megaphone and field glasses, and was used for many different purposes.

Fig. 17.—Tower Sixty Feet High From Which Transportation Was Superintended.

121

When brick were needed on any wall of any particular building, the foreman of that building would shout through his megaphone to the boy on the top of the tower. He, in turn, would shout through his megaphone to the foreman of transportation to carry the next turn of cart-loads to a certain place in a certain building.

122

The result of this method of handling on a large job proves that the saving more than justifies the cost of the runs. We believe this to be the most economical of all methods.

123

On city jobs where large quantities of brick, sand or such material are to be used, it is sometimes good practice to arrange a platform onto which the teams can drive and dump the carts down into the basement. Figure 18 shows such an arrangement. This platform will prevent blockades when there are numerous carts in line, will keep the street clean, and will save one handling of materials.

124

Figure 19 shows a good layout of plant for speed and economy. Two boom derricks were set up so as to pick up three

Fig. 18.—Sidewalk Traps for Rapid Handling of Material.

kinds of face brick, common brick, and three kinds of mortar, columns, girders, beams, flooring, cut stone, door and window frames, and masons' scaffolds. They were also used to raise each other up to the floor above. These two derricks did all the hoisting required on this building.

125

Figure 20 shows the construction of a seven-story hotel in San Francisco. On this job it was found to be cheapest to erect a 100-ft. derrick with a 95-ft. boom with a slewing rig on a tower at the second floor level of the building. This derrick hoisted the mortar and bricks from the basement, and the lumber and steel from the carts in the street, to the various floors. Where the wages of building laborers are very high, this is generally the most economical method of getting stock to bricklayers, provided there are also other heavy materials to be hoisted.

126

If elevators are used for handling stock, put a cleat on the elevator to stop the wheel barrow at the right place so that

Fig. 19.—Heavy Mill Construction Building in which all Material was Hoisted by Two Seventy-five Foot Boom Derricks.

neither the wheel nor handles will strike the floors as the elevator ascends and descends. If the barrow is to be hauled off backward, put the cleat so that the wheel will strike it, but if the barrow is taken off forward, make the cleat in two pieces and so located that the legs will strike them and wheel will go between them. Iron plates about ⅛-in. thick on the car under the legs of the barrow will save time in putting the barrow in exact place on car, as the barrow can be slid into place without lifting the handles.

127

When tending a bricklayer from above, as, for example on sewer work, requiring the lowering of brick from above, the brick must be piled in the exact manner shown in Fig. 21.

Fig. 20.—Derrick with Slewing Rig Erected on a Pedestal Two Stories High in San Francisco.

Bricks piled in this manner, i. e., courses of two headers alternating with three bull headers, will not fall out and hurt the man below.

128

If the bricks are piled up with the ordinary bond of an 8-in. pier, the bricks are sure to fall out occasionally, especially if they strike anything while being lowered.

129

It is generally economical to provide a board about 16 ins. square with two pieces of 1-in. board nailed to keep the brick from riding on the rope while the brick are being piled up. Two brick can be used for this purpose, but require more motions to operate.

130

The rope should be spliced on the hook. Old rope ¾-in.

in diameter is the best, as it will not twist so much when lowering, and it will lay closer to the brick. The hook should be on the side of the pile, near the top, but never at the top. Six feet of light chain between the hook and the rope will add to the durability and safety.

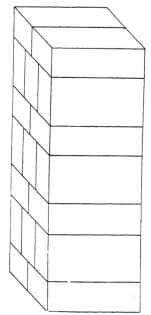

Fig. 21.—Bricks Piled for Lowering Into a Trench.

31

The handiest way to lower the load is to cover the leg above the knee with canvas or leather and let the rope draw across the top.

CHAPTER V.

SCAFFOLDS.

132

The kind of staging that can be used on a building to best advantage, and its location are matters that are handled differently on jobs where there are several contractors and on jobs where the entire work is done by different foremen under the same general contractor.

133

Figures 22 to 28 show a job that we built as general contractors in Montreal, where local conditions made a boom derrick with boom-slewing rig the most economical form of plant, because it conveyed the concrete of the footings, placed the stone of the foundations, set the cut stone, set the floor timbers and iron columns, and hoisted the brick and mortar. In fact it did all hoisting of all materials.

134

The bricks and mortar were hoisted in scale boxes, and were carried to the masons in hods.

135

This job is a good study in scaffolds, and illustrates the value of using for each case the kind of scaffold best adapted to secure the most speed and economy.

136

The Gilbreth scaffold was used for outside scaffold as high as it would reach, and outriggers were used above. On walls where no outside scaffolds were required, the Gilbreth scaffold was used for all inside scaffolding. Where outriggers were required, they were planked over both inside and outside. Trestle horses were used where there were many breaks in the wall. On the rear portion of the building that had no floors, the Boston scaffold was used.

137

In Fig. 23 note the kind of underflooring used, i. e., 2 x 4-in. stuff on edge. When this kind of flooring, in fact, when any kind of flooring over 2 ins. thick is used, at least one piece should be left out all around next to the wall until the roof is on. If this is not done the flooring may swell after a rain, and push the wall out of place. The missing piece can be put in after all possible swelling has taken place.

Fig. 22.—A Mill Building in Montreal. Observe the Dates.

138

When placing mortar boards, boxes or tubs for walls, composed mostly of piers, see Fig. 24, always have the mortar boxes and the brick so placed that the bricklayer can pick up the brick and mortar at the same time, and without taking a step.

Fig. 23.—Boom Derrick with Boom Slewing Rig Attached, which Hoisted all Material.

139
      If there is a left-handed bricklayer on the wall, the position of his mortar and brick should be reversed (see Fig. 24), so that he will not be handicapped in the athletic contest.

The above rule does not apply to blank walls, however. Where the wall has few openings, mortar boxes must be placed as shown in Fig. 25.

140

In Fig. 26 note the dump cart ready to dump the load of brick into the scale box to save one handling. Note the

Fig. 24.—First Floor on July 26, 1905.

method of increasing the height of the bricklayers' platform on the Gilbreth scaffold. Note the method of staying the window frames so that the stays will be out of the workmen's way until the frame is bricked in enough to permit removing the stay. Note the spreaders in the window frames to prevent the brick work from bowing in the frames. These spreaders are not made right in this figure. They are too long.

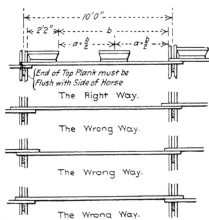

Fig. 25.—The Right and Wrong Way to Arrange Plank and Mortar Boxes on the Gilbreth Scaffold.

141

The spreaders should be made the exact length of the width of the frame, put in horizontally and tacked to a part of the frame that will not show the nail hole when the spreader is removed.

142

Make a run for hod carriers down on to the outside scaffold as shown in Fig. 27. Never permit the hod carriers to pass their hods to each other, if a run can be built that will not disturb the bricklayers.

143

Where there is a strong wooden floor, the best, safest, and cheapest form of fastening for the inside end of the outrigger is to spike a strip of 3 x 4-in. wooden joist on to the floor parallel to the outrigger and nail an 8 x 1-in. board to the strip and also to the outrigger in such a manner as not only to hold it up and down, but also to prevent the outrigger from rolling over sideways. See A in Fig. 28.

144

Outriggers may be attached to beams by a lashing twisted tight with the twister nailed to the outrigger. See B, in Fig. 28.

145

Where the weight is not too great, wooden centers for flat and segmental arches can be best supported by vertical pieces extending up from the sill below and held in place by

Fig. 26.—Second Floor on Aug. 2, 1905.

a spreader made slightly longer than the distance between the uprights.

146

If a few nails are necessary, use cut nails and do not drive them home. Leave at least ⅜-in. of the nail under the head out of the wood.

147

Nails driven clear in will not hold in green brick work one-half what they can be made to hold if driven as described

Fig. 27.—Third Floor on Aug. 7. 1905.

above. The spring of the board tends to pull them out after each blow of the hammer just enough to make them loose.

148

In Fig. 29 note the method of staying the back board. This is better than nailing to the outside plank.

149

The boom of the derrick being long enough to reach over all three walls, the various materials are piled by themselves.

150

The lumber is piled near the shed in the rear where the carpenters can work during the rain at framing timbers.

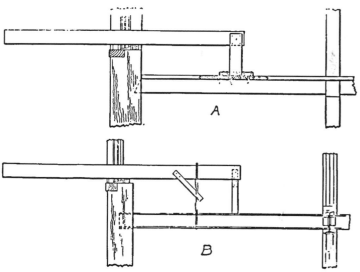

Fig. 28.—Methods of Staying Outriggers.

151

The iron work, cut stone and face brick are stacked up where needed, and the scale boxes are placed where the brick carts can dump directly into them.

152

While many brick may spill out of the scale box when the cart is dumped, one handling is saved on all brick that land in the box.

153

Note the way the outrigger scaffold is built around the corner. This method can be used when the corner piers are

not too large. When they are large it is sometimes convenient to leave a hole in the pier for an outrigger.

154

When it is necessary to build a staging up from the
ground on the outside of a building, as is generally the case

Fig. 29.—Fourth Floor on Aug. 14. 1905.

with digester houses of pulp mills, no better form of scaffold
has been devised where lumber is cheap, than what is called
the Boston scaffold, Fig. 30. It consists of vertical poles of
3 x 4-in. spruce, placed 8 ft. apart, and the width of five 10-in.
spruce planks plus 4 ins. (total 54 ins.) away from the face of
the wall.

155

Horizontal ledger boards 1 in. thick and at least 7 ins. wide are nailed on the inside of the poles with five 12-penny cut staging nails.

156

Putlogs of 3 x 4-in. spruce, or its equal, rest one end (without nailing), on the top of the ledger and against the pole, and the other end on the brick wall. The putlog may have to be notched (seldom over ½ in.) on its inside end, so

Fig. 30.—The Boston Scaffold.

that it will not be so high as to interfere with the bricklayers' line on the same course of brick. It must not extend into the wall more than one brick wide.

157

On top of the putlogs, lay the 2-in. plank. The plank should not be nearer the wall than 4 ins. unless there are men working underneath who might be hurt by falling brick.

158

Spring stays must be put on each ledger and each pole at every other junction of ledger and pole (see Fig. 30), unless there is an opportunity to stay in through the windows. Note the temporary stairway around the elevator tower.

159

When spring stays are used, the brick or wooden blocks must be placed as near the putlog hole as it is possible to have

Fig. 31.—Method of Staying Boston Scaffold.

it, without the spring stay breaking when it is bent down
to be nailed to the ledger. If the spring stay does not grip
the brickwork hard it is useless. The outer end of the spring
stay should be nailed with three 12-penny cut nails to the top
of the ledger and in a position hard against the pole.

160

Until a Boston staging is high enough to be stayed by
spring stays, it should be stayed by slanting braces nailed to
large stakes driven in the ground, as shown by Fig. 31. This
staging will then be so rigid that window frames may be
stayed to it.

161

Another effective method of staging and a much better
one than the spring stay, if it is also to be used to hold up
the staging plank, is to drive a wooden wedge hard in the putlog hole on top of the putlog, and drive two 12-penny nails
from the putlog into the pole.

162

Never drive a nail in a putlog except when it is used as a
stay.

163

A putlog should always rest against a pole and never on
the ledger away from the pole.

164

Be sure the pole rests on a piece of plank that will prevent
it from settling when the weight is put on it. Never stand a
pole on top of the ground without a board or piece of plank
under it.

165

If ledgers less than 8 ins. wide are used it is advisable
to nail cleats on the pole hard up under the bottom of the
ledger (see Fig. 31).

166

To splice a pole use two pieces of 1-in. board 4 ins. to 5
ins. wide and 2 ft. 6 ins. to 3 ft. long, with six 12-penny cut
nails in each board. That makes six nails in each pole. Put the
spliceboards so that they will be out of the way of the ledgers.

The ledger should be nailed against the pole and never
against the splice boards.

167

Nail the next ledger board on to the poles before the
masons start to work on the completed stage, as it forms a

Fig. 32.—Ledger Used as a Backboard.

good backboard to prevent the bricklayers and tenders falling over. (See Fig. 32.)

168

Never nail a ledger temporarily with less than five 12-penny cut staging nails. Many a staging has fallen because

Fig. 33.—Hanging Bracket for Supporting Scaffold Without Outrigger.

it was nailed temporarily, and then loaded before the rest of the nails were driven.

169

There is a great difference in the quality of staging nails. They should always be tested by being driven into wood to within 1 in. from the head of the nail, in which position they must stand bending forwards and backwards at right angles twice each way, without showing a sign of fracture.

170

The inside plank should always be turned bottom up and against the wall, or the brickwork will be spattered by the rain striking the mortar dust on the staging plank.

Fig. 34.—Outrigger for Inside as well as Outside Scaffolds.

171

Do not leave any planks on the staging below the place where bricklayers are working, because the dropping mortar will strike and spatter against the wall and cause expense to clean off.

172

A hole for an outrigger in the corner of the building can often be dispensed with by putting an outrigger over the roof and hanging down from it, as shown by Fig. 33.

173

When outriggers are used for exterior scaffolds and there are not many brick in the piers, and stays to window frames are numerous, it is sometimes good practice to use trestle horses on top of the outriggers, inside as well as outside, as shown by Fig. 34.

174

Figure 35 shows a method of building one story ahead of another without toothing. Toothing must be avoided wherever possible. Outriggers form a handy nailing for stays to window frames.

Fig. 35.—Outrigger Scaffolds for Two Different Stories on the Same Wall.

175

One of the many advantages of outriggers over any other kind of scaffold besides the fact that they leave no putlog holes, is that the frames and glass can be put in the windows, and the lower stories can be finished, as there are no staging stays in the way. This is shown by Fig. 36. Note the method of putting on the backboard to prevent bricks from falling off on the men underneath.

176

For exterior hanging scaffolds, where the bricks and mortar are passed out of the windows, there is no scaffold so safe and economical as the Murray Suspended Scaffold, Fig. 37. This scaffold is in two patterns: Where the winding drums are above and attached to the outrigger. Where the winding drums are below and attached to the scaffold platform.

Fig. 36.—Outrigger Scaffold on Type of Building Where it is the Most Desirable Form of Scaffold.

177

The latter pattern is the most desirable to use where the building is tall enough to warrant shifting the outriggers to a new and higher position, as the scaffold can be made fast to the lower outriggers, while the ends of the supporting cables can be carried up and attached to a new set of outriggers without stopping the bricklayers. Outriggers for this scaffold should be spaced 10 ft. apart.

178

Figures 38 to 40 show the hanging bar method of making temporary floors in a power station. There are often reasons for hanging scaffolds from above instead of building them up from below. Here is a case where speed of construction made it necessary to devise some form of scaffold hanging from above, so that the engine beds, machinery, piping, wiring

Fig. 37.—Murray Suspended Scaffold.

and finishing of the basement and first floors could be done while the exterior brick walls were being built. The scaffold hangers used consisted of 3 x ½-in. flat bars hooked over the lower chords of the trusses. These bars hung down in pairs and were punched 1 ft. on centers with 1-in. holes through which were placed ⅞-in. bolts with a ledger made of 3 x 12-in plank. These supported 6 x 6-in. putlogs 20 ft. long, and afforded temporary floors of exceedingly economical cost. Note here the two sets of double elevators. One set was located at each of the opposite ends of the boiler house and engine house of this power station.

179

The hanging bar form of scaffold permits finishing the work in the basement and first floor without having any scaffold in the way.

180

As soon as the walls were up to the ceilings, the planks were lowered, the hanging bars unhooked and lowered, and the entire interior was clear of all false work.

Fig. 28.—Arrangement of Plant for Speed of Erection of a Large Power Station.

Fig. 39.—Temporary Floors Hung from the Roof of a Large Power Station.

181

When starting bricklayers at the ground level always lay down plank for the bricklayers to stand on. If they have a proper place to stand upon, they will lay more than enough brick to compensate for the extra labor of laying down the plank.

Fig. 40.—Temporary Floors Near Under Side of Roof Trusses.

182

When there is a wooden floor, no preparation for a scaffold is required.

183

Have all stagings inspected by an intelligent bricklayer before they are used. Have him say in the presence of wit-

Fig. 41.—Method of Covering Masonry Projections.

Fig. 42.—Stays So Made for Window Frames that They Will Not Retard
Bricklayers.

nesses that the staging is right, before it is used. We shall
then know that we have done everything in our power to make
the stagings secure, and to prevent accident.

184

  Cover all projections of brick work as shown in Fig. 41,
so that the falling bricks will do no harm, and so that the
falling mortar will bound off instead of pile up. Be sure
that the covering is so stayed that it will not blow off in a
high wind and injure the public in the street.

185

  All window frames should be handled and set by carpenters.

Fig. 43.—Correct Way of Staying and Lining Frames.

Fig. 44.—Type of Old-Fashioned Staging Horse.

186

The carpenter should nail the stays to the frames in such a place and manner that they will interfere in the least with the bricklayers. (See Fig. 42.)

Fig. 45.—Type of Trestle Horse Staging.

187

As soon as the frame is in its proper place and stayed, the carpenter should mark with a pencil on the sill around the end of the frame, so that it will be readily detected if the frame has been moved out of place.

188

If the carpenter fails to pile up brick on the frame to weight it down, the bricklayer should pile up enough brick upon it to keep it in place until it is bricked in.

189

Window frames will seldom be plumb and straight unless the foreman bricklayer and foreman carpenter co-operate. If the frames are near together, a continuous piece of joist tacked onto the frames will help to keep them in place while they are being bricked in. (See Fig. 43.)

190

It is the duty of every bricklayer who is in a position to sight along the window frames to notify the foreman bricklayer if the frames do not appear to be all in line as well as all plumb.

191

Put a spreader in all tall frames, or the bricks will surely push and bow the sides in.

192

Figure 44 shows a type of scaffold horse that is used in many parts of America, but should never be permitted on our work.

193

All inside scaffold should either be the Gilbreth scaffold or else the standard 4-legged trestle horse.

194

Compare the method shown by Fig. 45 with that of the Gilbreth scaffold. Bricklayers cannot be expected to lay as many brick when working in this position as when standing up between a wall and a stock platform, each 2 ft. high, where one can stand as if working at a bench.

CHAPTER VI.
GILBRETH SCAFFOLD, HOD TYPE.

The Gilbreth scaffold accomplishes two things:

(A)   Helps make better workmanship.

(B)   Helps make more economical workmanship.

(A)   It helps make better workmanship for four reasons:

(a)   It keeps the bricklayer at a constant height, where he can lay his brick with the most precision.

(b)   It enables the bricklayer, even on overhand work, to keep the wall backed up solid, i. e., the entire width of the wall.

(c)   The wall being always backed up solid, it can be properly protected from rain and sleet by simply putting the staging planks on it.

(d)   The bricks can be best bedded because the work is always at a height that the bricklayer can best handle the brick and mortar.

(B)   The Gilbreth scaffold makes more economical work because:

(a)   It keeps the bricklayer at a constant height, and that height is such that no stooping over is required to pick up brick and mortar, and no stooping nor reaching is required to lay brick and mortar.

(b)   It reduces the number and the length of motions.

(c)   It does away with the bad practice of backing up 4 ins. or 8 ins. against the overhand or exterior face tier (see Fig. 46), and permits all filling of each course to be done the entire width of the wall at once.   (See Fig. 47.)

(d)   The bricklayers' platform is clear at all times, and there are no bricks underfoot.

(e)   The bricklayer is out of the tender's way and the tender does not interrupt the bricklayer as he passes back and forth.

(f)   When ordinary scaffolds are used, the wall is generally built 5 ft. above the floor. See Fig. 46. When the Gilbreth scaffold is used, the wall is not built more than 3 ft. 8 ins. above the floor. This 1 ft. 4 ins. when built from the floor is where the most reaching has to be done and consequently that extra time and labor is all saved by set-

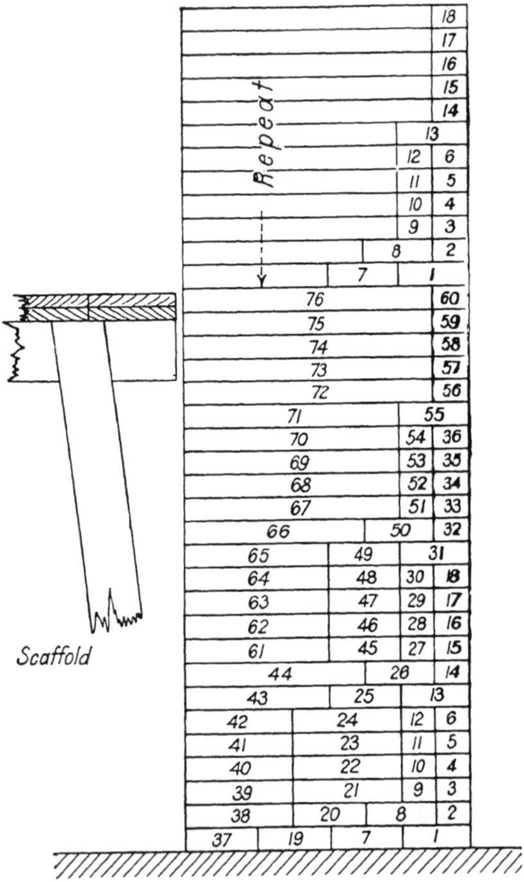

Fig. 46.—Sequence in Which Courses and Tiers are Built Overhand from a Trestle Horse Scaffold.

ting up the Gilbreth scaffold when the wall is 3 ft. 8 ins. above the floor.

(g) When the Gilbreth scaffold is once set up, and it takes no longer to set it up than it does to set up a trestle

horse scaffold, there is no more worry, trouble or stage
building for that wall in that story. No time is lost in shifting men from wall to wall. No athletic speed contests are
broken up. This last feature alone is of great importance.

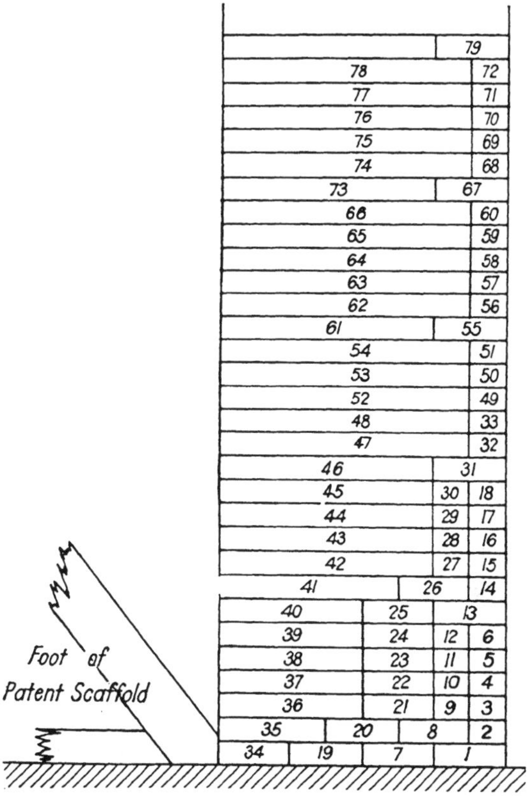

Fig. 47.—Sequence in Which Courses and Tiers are Built Overhand from
the Gilbreth Scaffold, Hod Type.

196

When the Gilbreth scaffold is to be used and there is
not a wooden floor to nail it to, lay down three rows of plank
as shown by Fig. 48.

197

Put the inside plank hard against the wall.

198

Put the outside plank out to the end of the bottom piece of
the horse.

Fig. 48.—Temporary Floor for the Gilbreth Scaffold.

199

Put the center plank where it will come exactly under the uprights of the horse.

200

Use no nails except in the end of the stay rods.

201

Do not build the wall more than 3 ft. 8 ins. above the floor before setting up the scaffold. Take advantage of the

Fig. 49.—Setting Up Gilbreth Scaffold.

economies of the Gilbreth scaffold in every course more than 3 ft. 8 ins. above the floor.

202

Superintendents must see that the story poles that they give to the men on the leads have marked on them, at the height of 3 ft. 8 ins. that the scaffold must be set up as soon as the wall is 3 ft. 8 ins. above the floor. Otherwise the men will build the wall 5 ft. high in accordance with the old fashioned custom.

203

Stand the horses up 10 ft. 6 ins. apart, with the red side against the wall. This means that the shorter end of the foot

Fig. 50.—Hooking One Stay Rod to Each Horse.

of this scaffold horse must actually touch the wall. This will give exactly 1 in. clearance to the arm supporting the bricklayers' platform.

204

This scaffold is designed to very close measurements, and is the result of many years of actual practice. These rules must be carried out literally. (See Fig. 49.)

205

Hook one stay rod in an eye bolt of each horse. There is an eye bolt on each side of each horse and it does not matter to which side the rod is hooked. Only one rod is needed to each horse. See Fig. 50.

206

Drive two 10-penny cut staging nails into the holes in the lower end of the stay rod.

207

Never, under any circumstances, drive any nails into the horse. There has never been an occasion where nails in the horse were necessary.

208

The first position of the frame in the horse is as shown by Fig. 50, resting on the foot of the horse, not resting on a pin in the horse. This is important.

Fig. 51.—Tenders' Platform Two Plank Wide.

209

Use no planks less than 2 x 10 ins. Make the brick-layers' platform two planks wide. Make the stock platform three planks wide. Make the tenders' platform one plank wide when it is first set up, and two planks wide when the staging has been jacked up one or two notches. (See Fig. 51.) Do not set up the frame in the horse high enough at first to permit two planks on the tenders' platform.

210

Do not use a plank on the stock platform long enough to rest on more than two horses.

Fig. 52.—A Large Number of Runs to Save Time and Money.

211

Mortar boxes should not be over 4 ft. apart. For some inexplicable reason, foremen do not realize the benefits accruing from having boxes near together. Even the bricklayers themselves, when it is left to them, will place their boxes too far apart. On our work, boxes must never be more than 4 ft. apart, and on the Gilbreth scaffold, the mortar boxes and the plank must be arranged as shown in Fig. 25.

212

Watch carefully the circuit the tenders make when leaving the stock pile, dumping and returning to the stock pile. It is

Fig. 53.—Long Cleated Runs to Save Ladders.

always economical to make return trips by special short cut runs.

213

When using wheelbarrows on tenders' run, be sure that the planks are laid so that the tenders will wheel down upon the next plank, not up onto the next plank.

214

Every time each hodcarrier passes a bricklayer on the same stage, he interferes with the speed of the bricklayer. Many different runs, Fig. 52, will enable the hodcarrier to go and come the shortest way and pass the fewest bricklayers.

215

In many localities it is hard to procure skilled hodcarriers who can climb a ladder with a hodful of brick or mortar. Therefore, oftentimes a long cleated run, Fig. 53, will do away with the necessity of a ladder. For these runs use plank 3 or 4 ins. thick, so that they will not be springy and will not require props. If cleats are used, they should not be more than 1 in. thick, and they should be spaced at even distances.

216

It is not possible to make a rule covering all cases, that will state how far apart the cleats should be. The distance apart depends on the kind of men and the slant of the run, and is a matter of sufficient importance to warrant experimenting under the local conditions. All runs should be wide enough for men going up and down to pass one another.

217

The boss of the wall must notify the foreman if stock is not coming fast enough, also if the mortar is too rich, too sandy or not right in any way.

218

It is good practice to keep the floor around the elevator one story higher than the rest of the building. This will enable the tenders to go down to the high stagings with their load and to come up from the stagings empty.

219

Have the man who jacks up the scaffold stand on the tenders' platform, Fig. 54, not on the bricklayers' platform.

220

Do not let him pull out a pin and put it in the hole above.

See that he has a spare pin and that he puts the spare pin in place in the upper hole, lowers the frame down on it,

Fig. 54.—Jacking Up the Scaffold While the Men are Working on It.

and then carries the lower pin to the next horse and uses it as a spare pin. In other words, there must always be a pin in the horse as a double safeguard.

221

To hold a frame in place, the pin should go in under

Fig. 55.—Location of the Tenders' Platform.

the upper member of the frame. It should never go through
the iron hanger of the frame, nor in any other hole.

222

Never jack the stock platform higher than the inside 4 ins.
of the wall, as otherwise it will increase the length of motion
for each brick.

223

Note in Fig. 55 that the tenders' platform is located at
such a distance below the stock platform as to enable the
tender to dump his hod with the least effort and with no
more climbing than is absolutely necessary.

Fig. 56.—Method of Adding More Braces Without Any Nails.

224

When tall horses are used, it is sometimes desirable to prevent them from swaying. In this case planks
butting against the foot of one horse and leaning against the
uprights of another horse (see Fig. 56) will do the work
perfectly. Use no nails.

225

Bracing two or three horses in this manner will generally stay all of one line of horses.

226

Note how the horse is used to hold up the run.

227

Note that the tender in the center of Fig. 56
is loading his hod the right way. Note that he is straddling
the shank of the hod, and is picking up the brick with both
hands at the same time. Foremen should see that all brick
hods are filled this way.

228

In places where hodcarriers are hard to procure, unskilled laborers will have to be taught the arrangement of the brick in the hods, which will differ with the size, and with the shape of the men and their shoulders. Regardless of the arrangement of the brick, the foreman should know how many brick each man is carrying. Unless there is some special reason to the contrary, they should all be made to carry the same number of brick.

229

The Gilbreth scaffold is particuarly profitable on overhand work, because it maintains the bricklayer always at the right height to do the overhand face work in the best manner and to the best advantage. (See Fig. 57.)

230

Fig. 58 shows that a bricklayer can lay more brick in the middle of the wall when he is filling in between the outside tiers than when he is backing up the overhand 4 ins. with a backing 4 ins. or 8 ins. thick, as more precision of workmanship and more motions are required in the latter case.

231

It is also obvious that the best wall is one that is backed up solid at least every header high.

232

Figure 46 shows a wall built overhand in the usual manner, from the old-fashioned type of scaffold. It is built 5 ft. to 6 ft. high. It is a slow, expensive, difficult process to build any portion of the wall more than 3 ft. 8 ins. above the place on which the mason stands. This same method and process must be repeated in exactly the same manner on every "staging high" of the wall. The men must be shifted to some other place every time the staging is built. Only those who have specially timed it realize the great amount of time actually lost by shifting men about.

233

To overcome the above disadvantages, if for no other reason, one of the two types of the Gilbreth scaffold must be used on all walls wherever possible, as it reduces all these disadvantages to a minimum. See Fig. 47.

234

The height above the floor that the wall should be when the scaffold should be set up for the greatest economy is a

Fig. 57.—Advantages of the Gilbreth Scaffold for Overhand Work.

much discussed question among bricklayers. On our work the following rules shall be followed:

235

For walls laid overhand from the Gilbreth scaffold—

(a) If the wall is five tiers of brick thick, or less, set up the scaffold when the wall is 3 ft. 8 ins. above the floor.

(b) Set up the scaffold at one course less than 3 ft. 8 ins. in height for every tier that the wall is over five tiers thick.

236

For walls laid overhand from trestle horse staging—

(a) Divide the height of the wall into even "stagings high" of not less than 4 ft. nor more than 6 ft. high. If the

Fig. 58.—Old Method of Backing Up Solid After the Overhand Face Has Been Built Staging High.

wall is thick, it will be more economical to make and raise stagings oftener, so as to have less reaching for the bricklayers. If the wall is two, three or even four tiers of brick thick, it is generally more economical to have the bricklayers lay even 6 ft. high than to raise an extra tier of scaffold horses and to lose the time that is always lost when moving men from one staging to another.

237

Stagings for backing up only—

(a) Set up the Gilbreth scaffold when the wall is 3 ft. 8 ins. high above the floor.

(b) Set up 4-legged trestle horses when wall is 4 ft. 6 ins. to 6 ft. high.

Fig. 59.—Stagings on Both Sides for the Greatest Speed.

238

Figure 46 shows by the numbers the sequence in which the various portions of a brick wall shall be built when constructed overhand from a trestle horse staging.

239

No sequence will be permitted that changes the position of the headers from the position in which they would be placed if laid from both sides.

240

Fig. 47 and Figs. 65, 66, 67 and 68 show the sequence in which the various portions of the same wall would be built from the Gilbreth scaffold. These two diagrams show conclusively the great economy gained with the Gilbreth scaf-

Fig. 60.—Taking Down the Gilbreth Scaffold.

fold, if it is used in a manner that takes advantage of all of its possibilities. The greatest advantage illustrated by these figures is that the brick may be laid in large level areas instead of requiring to be backed up one or two tiers at a time, to get on the headers on the overhand face.

241

A small amount of motion study will show the advantage of not being obliged to stoop to get the stock for that portion of the wall that is between the height of 3 ft. 2 ins. to 5 ft. above the floor; and, in time, analysis will show that no allowance will have to be made for the time that men are shifted from one wall to another while the staging is being raised.

242

Where the greatest speed is required, set up this scaffold

on both sides of the wall. With the scaffold so set up almost
any wall can be built a story high in one day. See Fig. 59.

243

    When taking down the horses of the Gilbreth scaffold—

    (a) Clean off all stock, preferably carrying it to where
it will be used, in order to save one handling. See Fig. 60.

Fig. 61.—Sliding Down the Mortar Boxes.

    (b) Slide down the mortar boxes, as shown in Fig. 61.

    (c) Take down the planks, one at a time.

    (d) Pull up the stay rods from the floor and then un-
hook them from the horse.

    (e) Lay the horses down on the floor.

    (f) Slide the frames down to the bottom of the horse
after the horses lie flat.

244

    Do not drop the frames down while the horses are stand-
ing, or you will surely break the bottom piece of the horse.

CHAPTER VII.

GILBRETH SCAFFOLD, PACKET TYPE.

245

The Gilbreth scaffold, packet type, is a modification of the Gilbreth scaffold especially adapted to the packet method and the fountain trowel. It is so designed that it may be set up and put in operation before the wall is built any higher than the floor on which the bricklayer stands.

246

There are several reasons for setting the scaffold up at once, as soon as the wall is level with the top of the floor.

247

First, it is easier, and quicker, consequently more economical, for the tender to unload the packs of brick from the wheelbarrow to the stock platform, than to lower them down to the floor.

248

Second, it is easier and quicker, consequently more economical, for the bricklayers to take their bricks from the stock platform than to stoop over to the floor for them. Furthermore, with this scaffold they can throw the mortar from the mortar boxes into the wall without stooping either at mortar box or at wall.

249

Set up the scaffold when the wall is as high as the floor on which the mason stands.

250

Place the horses not less than 10 ft. nor more than 10 ft. 6 ins. apart on centers. It is important that this distance does not vary more than the limits given, or it will seriously interfere with necessary working conditions explained later.

251

Place the inside end of the footpiece of the horse against the wall (see Fig. 62). This is important, because if the bottom piece is against the wall it will make the support for the bricklayers' platform clear the wall by exactly 1 in.

252

Hook one stay rod onto each horse, and nail it to the floor with two 10-penny cut nails. One iron stay is enough for each horse.

253

Put two planks and a center board on the stock plat-

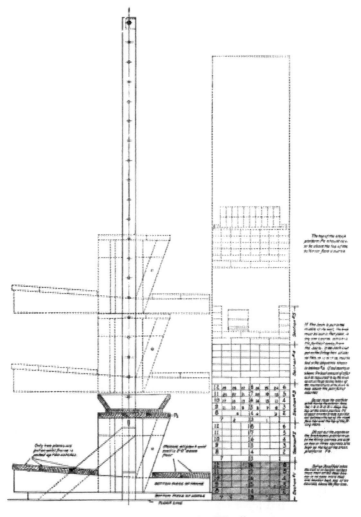

Fig. 62.—The Packet Type for Wheelbarrows.

form. Lay the plank as shown in Fig. 62. Shove the "center plank" hard against the uprights of the right-hand horse, i. e., the one nearest the hauling lead of the wall. The hole left between the other end of the center plank and the left-hand horse is covered by the left-hand mortar box.

254

Shove this box hard against the left-hand horse, and place another mortar box equidistant between this first or left-hand mortar box, and the right-hand horse. This will leave two equal spaces about 2 ft. 9 ins. long for the brick boards.

255

Bolt two pieces of wood 2 ft. 5 ins. long crosswise into the stock platform for a track under the brick packets. The tender can lift his pack of brick from the wheelbarrow

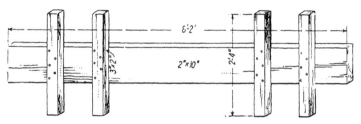

Fig. 63.—Center Board for Stock Platform, Packet Type.

to the stock platform easily, if he does not have to reach in too far over the edge of the stock platform. Therefore, the stock platform must have two tracks for each brick space. They must be located in every case exactly as shown in Fig. 63. These permit dragging and pushing the brick packets in toward the bricklayer.

256

Provide three planks for the wheelers' platform, but do not put them in place until the frame has been jacked up one or two notches, or until the planks on the wheelers' platform will clear the braces on the foot of the horse. If the planks are of a kind of wood that bends too much, one or two cleats on the under side of the platform will remedy this.

257

At the first position of the scaffold, the top of the stock platform is 2 ft. 11 ins. above the floor on which the scaffold, and also the bricklayer, stands.

258

Do not lay any plank on the bricklayer's platform until the wall has been built at least 2 ft. high above the floor.

259

As soon as the wall has been built 2 ft. high above the floor two plank must be laid on the bricklayer's platform. The bricklayer is then in a still better position, and at a more convenient height to work at his greatest speed.

260

Besides all these advantageous considerations, the gang has not been interrupted nor interfered with, and

Fig. 64.—The Gilbreth Scaffold, Packet Type, for Wheelbarrows.

no time has been lost shifting men from the wall and back again while a scaffold has been set up. This last consideration means a large saving of time on a large gang, and overcomes the possible disadvantage of not having the same man continue building the same corner, or angle, or plumb spot, from bottom to top of the wall, with the consequent divided responsibility for inaccurate work.

261

The stock platform must be kept at a height of not more than 8 ins. (the spacing between holes in the horses), below, and never above the top course on the inside face tier of the wall. See Fig. 64.

262
    This is a matter often overlooked by the busy foreman.
*Next to seeing that the line is hauled the instant that the last
brick is laid out, the jacking of the scaffold and the maintaining
of it at the right height are the most important features to watch
for the greatest economy.*

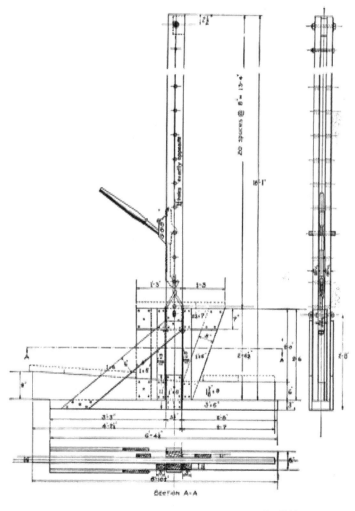

Fig. 65.—Dimensions of Gilbreth Packet Type Scaffold.

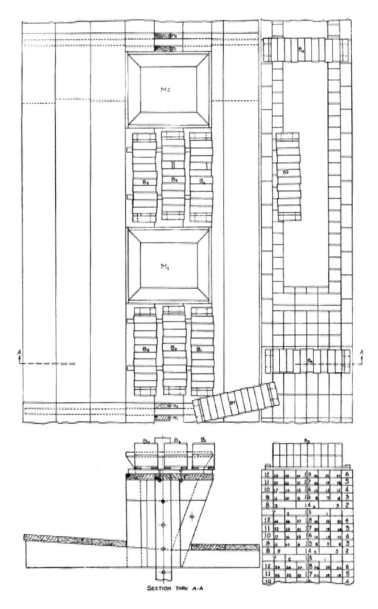

Fig. 66.—Location of Packs for Shortening Distance of Transporting Brick to Wall.

263

The average foreman underrates the necessity of having the scaffold at exactly the right height, because he sees that the bricklayer can stoop and bend and yet lay brick almost as fast as when he does not bend or stoop at all. Not being over-worried about the bricklayer's comfort, so long as he is laying brick fast, he forgets that men working at manual work, like bricklaying, cannot keep up the work every instant, and that, therefore, the percentage of rest absolutely required by such men must be greater than that of the men who are put to no exertion not absolutely necessary to laying a brick.

264

It is not enough for the foreman to put a laborer jacking up the scaffold who knows how to jack it up. He must instruct the laborer until he understands that the scaffold is to follow and to be kept at the height of the inside face tier, not the outside face tier; that the stock platform is never to be above the inside face tier, and that when the top of the filling tiers is below that of the inside face tier, the stock platform must be kept at a level half way between the grade of the top of the filling tiers and the inside face tier.

265

When the filling tiers are as high as the inside face tier, the stock platform is to be kept level with the top of the inside face tier.

266

The height of the outside face tier has nothing to do with the height of the stock platform on this scaffold.

267

When laying the outside face tier, the stock platform must be as nearly level with the top of the inside face tier as possible.

268

The bricklayer can then transfer the packs of brick from the stock platform to the wall without lifting the pack more than an inch or so, and if the stock platform is maintained at the relative heights here described, the bricklayer can do this transferring without stooping and with no bending of the back. Therefore, in reality, to transfer the pack of brick in a level plane from the stock platform to the wall requires no more work to be done than the stooping of the body and the straightening it up again. See Figs. 65, 66, 67, 68.

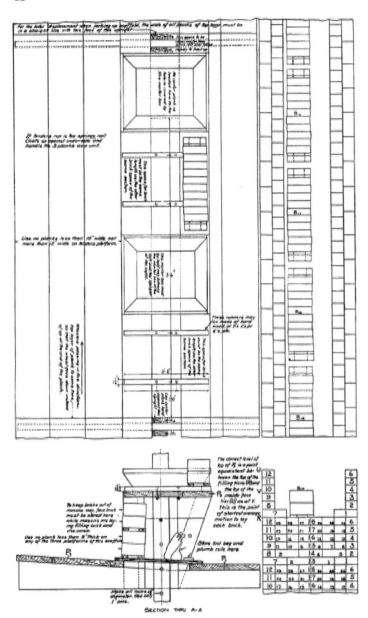

Fig. 67.—General Arrangement of Packet Type.

269

A great advantage of this type of scaffold is the fact that it enables the bricklayer to back up his wall solid every "header high," starting from the floor level itself.

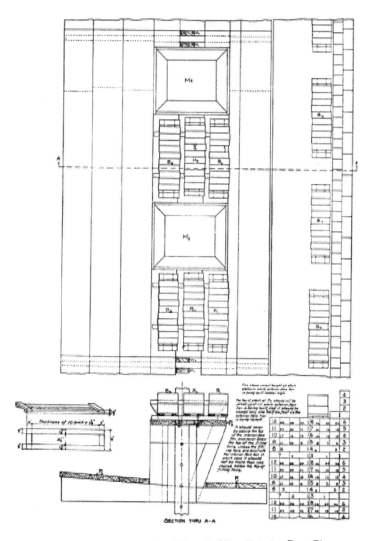

Fig. 68.—Location of Packs When Building Exterior Face Tier.

270

It is obvious that a bricklayer can lay more brick, lay them more evenly, bed them better, and shove the joints easier when he is backing up solid, i. e., the full width of the wall, than when he is backing up one or two tiers thick, to get on a header, so that the exterior face tier can be built up staging high.

271

This is the only form of scaffold that permits backing up the wall solid from the floor line up to the ceiling.

CHAPTER VIII.

THE GILBRETH PACKET SYSTEM.

272

The Gilbreth packet system consists of conveying bricks upon packets from the pile in the street to the top of the wall.

273

If the bricks are brought "packed," i. e., side by side, on edge, to the job, then they are to be unloaded and placed upon packets regardless of whether or not they are to be used at once, except in the case of brick that are to be culled before they are used.

274

If the brick are brought "loose," that is thrown in, they are to be dumped out, if the car or cart is provided with dumping means. But if they are not dumped out, they must be put upon packets, carried out or wheeled out, whichever is the cheaper, and stacked up on the packets until ready to be used.

275

The packets shall be made of two pieces laid lengthwise, and so spaced that the outside edges of the packet are spaced exactly the length of the average brick to be carried. The space between the two pieces shall be wide enough to permit room for the men's fingers to clear without jamming. See Figs. 68 and 69.

276

The lengthwise pieces shall be held in place by one crosswise piece at each end, that shall be so spaced that the distance in the clear between them shall be $1\frac{1}{2}$ ins. greater than the length of the lower layer of brick on the packet. Round off all corners of the end pieces where the hands rub.

277

The method of handling the packs from the stock pile to the hoisting apparatus, and from the hoisting apparatus to the mason, depends upon circumstances. Sometimes it is cheaper to have laborers carry them in their hands (see Figs. 70 and 71), sometimes skeleton wheelbarrows holding three or

Fig. 69.—Arrangement of Mortar Box and Packs to Obviate Stooping.

four boards are the cheapest (Fig. 72), and on long runs it is
sometimes most economical to pile the packs of brick into the
carts. (See Figs. 73 and 74.)

278

In buildings divided by brick walls into small areas, it is
often difficult to build long sloping runs up for wheelbarrows.
When a run cannot be arranged down from the floor above,

Fig. 70.—Unloading a Freight Car with a Gravity Conveyor.

Fig. 71.—Handling Packs on a Gravity Conveyor.

packs can be passed from tender to tender, from floor to stock platform, by having a few stagings 3 ft. high one above another.

279

When the packs reach the bricklayers' platform, they are shoved over on the tracks by the tenders toward

Fig. 72.—Wheeling Packs.

Fig. 73.—Loading Carts from Gravity Conveyor.

the bricklayer. It is a very simple matter to pick up 90 lbs. when the lift is straight up, but it is a very difficult matter to pick up 90 lbs. when the lift is not straight up. Therefore the bricklayer must have his brick put as close to the inside edge of the stock platform as possible, so that he can lift his load in the easiest manner.

Fig. 74.—Preventing Packs from Spilling in a Wagon.

280

The number of brick that shall be piled upon a packet varies in different localities on account of two factors (a) the size and weight of the particular brick used, and (b) the quality of the laborers obtainable.

281

The weight of the brick that should be put on a packet should, with the weight of the packet, be as nearly 90 lbs. as is possible, with an even number of brick. This is the weight which a first-class, high-priced laborer can handle to the best advantage. With inferior, low-priced laborers, the number of brick must be reduced so that the weight will be lessened in proportion to their strength.

282

The strength of the laborers and not the strength of the bricklayers is to be the controlling factor to determine the weight of the load on each packet.

283

Many small men who have been rated as first-class bricklayers can work rapidly with a light load of, say a brick in each hand. They could not possibly stand the strain of transferring the packs of brick from the scaffold in a horizontal plane to the wall.

284

Other bricklayers, who are able to handle 90 lbs. with ease, would much prefer to lift a pack containing 20 brick from the scaffold in a horizontal plane to the wall than to make ten trips with a brick in each hand each trip.

285

Any bricklayer who is not able to transfer the packet with its regular full load from the stock platform to the wall will be obliged to take off a few brick from each pack in the old method until the remaining load on the packet has been reduced to a point where he can handle the weight comfortably and thrive under the continuous exercise of it. Of course he cannot earn as high wages as the man who can handle the larger load.

286

The bricklayer must place the pack on the wall in the location that will give the shortest possible distance through which to carry each brick from the packet to its final resting place on the mortar.

287

The bricklayers do not realize the importance of this, and must be constantly instructed to place the pack as near the place where the brick are to be laid as possible, even with the packet method. Careful packing of the pack so as to save all the motion possible will diminish the distance that a bricklayer's hand travels from a quarter to a half mile of distance per day.

288

When laying the outside face tier, the bricklayer will convey the packet to the top of the wall. If it is a thin wall, he will lay the outside face tier while standing on the bricklayer's platform.

289

If it is a thick wall, he will lay the outside face tier while standing on the top of the filling tiers.

290

After the outside face tier has been built up to the bottom of the next header, the inside face tier will be built up several courses above the filling tier. Then the filling tiers will be laid.

291

The advantages of the packet method are to be seen all through the process of bricklaying. The brick themselves will be kept in better condition. Unloading brick from a hod or wheelbarrow is sure to make more or less bats, and many chipped brick. With the packet method it is possible to have the brick arrive at the scaffold without a chip out of them. This means the saving of the time needed to discard bats and to select unchipped brick fit to lay to the line. It also, of course, means a better looking wall.

292

Another saving of time occurs in that the bricklayer does not have to separate his stock into that suitable for face work, and that fit only for filling. With the packet method it is a simple matter to put the face brick on some packets, and the filling brick on others, and to keep them separate until the bricklayer needs them.

293

The packet method also gives the bricklayer something to do when he would otherwise be idle, i. e., when he is waiting for the line to be raised, waiting for mortar, etc.

294

All practical bricklayers know that the best results cannot be obtained if the men lay brick above the line. Nearly all well-governed unions make it a rule to lay no brick above the line, and never to slack out a line until it is all laid out. While these are splendid rules, and make for justice to the men and better walls for the owner of the building, the man who pays those bricklayers begrudges the time that these rules sometimes cause to be wasted.

295

With the packet method, the bricklayers can take packs of brick off the stock platform and place them on the wall in the position that will be the nearest possible to the place where they are to be laid. This not only affords the bricklayer a better chance to earn his bonus, but it also leaves just so much more room on the stock platform and just so much less chance of the bricklayer running out of stock in case, for any reason, the tenders are delayed; both great aids toward speed and economy.

296

Some idea of the economy of work done by a bricklayer with packets on this scaffold is apparent when one considers that a bricklayer, lifting 90 lbs. without stooping, lifts about 90 lbs. for 20 brick, while on the old-fashioned scaffolds and methods he bends over and lifts all of his body above the waist and 9 lbs. every time he lifts 2 brick: this is, about 109 lbs. for 2 brick on the old-fashioned method.

CHAPTER IX.
TALL CHIMNEYS.

297

The problem of getting stock of the kind that the brick-layers need to the top of the chimney, as fast as they can use it has been successfully worked out. In the future, any chimney large enough to have an elevator must also have a track running in through the temporary stock opening and a track on the platform of the elevator for cars of stock. See Fig. 75.

298

The track must be slightly down grade to the elevator (about 1 in. in 5 ft.) so that loaded cars will run down hill and the empty cars up hill.

299

This track must be long enough to serve many loads of brick hauled and dumped from carts alongside the track, and also to serve the mortar beds.

300

The track should have branches but no turn-outs, as no loaded cars need pass each other.

301

The empty car is so light that it can be picked up and carried past the full car which has been pushed in close to the elevator and is waiting for it to come down.

Fig. 75.—Track Through Stock Opening Into Elevator.

302

When the elevator comes down the next loaded car should be inside the chimney ready to push onto the platform. A small school slate attached to the top of the elevator car will tell whether outside brick or mortar or inside brick or mortar is wanted on the three or four following cars.

303

The empty car should be taken off and carried out, and the full car pushed on.

304

The cars are loaded with packets carrying two layers of bricks on edge, generally about 20 bricks on a packet.

305

The mortar should be handled in metal pails or in fountain trowels. These are stacked up four pails high, each pail having a piece of board over it and one under it to steady the pile of pails.

306

Nail some old rubber hose with upward loops about 4 ins. high underneath the bottom of the car for the car to strike. It will prolong the life of the car, and so may prevent delays.

307

It is never good practice to have the elevator rope pass through the opening where the bricks and mortar are carried in, as the rope is in the way of the men.

308

The scaffold at the top of the chimney must be built specially from individual designs, after the design of the Gilbreth patent scaffold, made by our office to suit the size of the chimney.

309

The stock platform must be located exactly 2 ft. above the place where the mason stands. It must have a back board at least 5 ins. high and ⅞-in. thick.

310

The masons' platform must fit the inside of the flue with not over 1 in. play. The masons' platform must extend in under the stock platform at least 8 ins., and more if possible.

311

The edge of the masons' platform must be boarded up tight to the stock platform, so that nothing can drop on the men loading the elevator below.

312

The entire scaffold must be nailed or bolted fast together, but loose on the uprights.

313

The uprights must be not less than 4 x 6 ins., nor more than 6 x 6 ins., with the 6-in. face for the guide surface.

314

These two uprights must be bored with 1-in. holes, exactly 8 ins. apart on centers, for holding the lifting jacks. These uprights must be bored before being erected and must extend 4 ins. beyond the center of the last hole.

315

The well opening must be covered with the Bowler automatic platform lid, made to prevent bricks from falling down the flue and strong enough to sustain two tenders.

316

The stock platform must be maintained about level with the inside 4 ins. of the brickwork, so that there will be no stooping nor long reaching.

317

This outfit will cost about $10 more to build than the old method. With it bricklayers should average twice to three times the usual number of brick per day, laying shove joints. We have had them lay over 4.200 brick with shove joints on first-class jointed work in 8 hours.

318

Chimneys having no ornamental work on the exterior need no outside scaffold.

319

The best results in round chimneys are obtained by having no brick headers below the head of the chimney. Get permission from the engineer or architect to use galvanized wire ties instead of brick headers for bonding the outside tier of round chimneys.

320

Examine the mortar of the inside face of the chimney from the top to bottom every two days, to see that it is setting fast enough to carry the work being built above it.

321

Build an outside protection over the men at the bottom of the chimney. Make this strong enough to stop a brick falling from the top of the chimney. Test it.

Fig. 76.—A Round Chimney 255 Feet High (34 Ft. Higher Than Bunker Hill
Monument).

Fig. 77.—Concrete Foundation and Template Ready for Building Brick Chimney.

322

Figs. 76 to 83 show our typical methods of building a tall chimney. It was built in record time. The top 100 ft. were built in 14 days. All of our foremen must use the methods here shown on future chimneys.

Fig. 78.—Beginning the Erection of the Exterior Scaffold.

323

Set a wooden plug, with a small headed wire nail driven home into the top of it, in the center of the top surface of the foundation.

324

Build a substantial templet of the exact size of the bottom of the chimney. This templet will more than pay for its cost by saving time in checking up offsets. (Fig. 77.)

325

Superintendents must pick out only those bricklayers for tall chimney work whom they know personally to be thoroughly reliable, good mechanics—men accustomed to laying bricks with shove joints and who can be absolutely depended upon to do so. No other kind of work except shove joints is

Fig. 79.—Back Filling Completed, and Everything Ready for Quick Construction.

allowed, and all joints must be absolutely full of mortar. Make the best man the working foreman.

326

The core and the shell of a chimney must never be connected in any way, except at the base of the chimney. Here the bases of both must be entirely of headers to the top of the offsets of the footings. See Fig. 78.

327

Tend the masons with hods or packets and pails, Figs. 78 and 79, until the chimney is two stagings, or about 10 ft. high. Then the inside elevator should be installed.

328

The fastest method where conditions will permit is, of course, to have the tenders place the packs containing the

Fig. 80.—A Boston Scaffold for the Erection of the Exterior Face Tier. (Note the hole left in the front side of the chimney for the track to the elevator.)

Fig. 81.—At This Height the Exterior Scaffold Was Discontinued.

brick near the inside edge of the stock platform and parallel
with the edge so that the masons can handle the packs
to the top of the wall, where they can pick them up
and lay them with the least possible amount of reaching. This
will permit much faster work and will enable the bricklayer to
earn higher wages. For full description of the packet method
see Pages 85-91.

329

If there is any fancy brickwork on the exterior, or if
a different kind of brick or mortar is used on the exterior
than the interior, it is generally advisable to build an outside
scaffold at least 25 ft. high. (See Fig. 80.) If there are any such
fancy offsets from octagonal to round, as there are on this
chimney, the scaffold must be built up higher. On the largest
of chimneys there is not any too much room inside for men
and stock. If this space were to be filled with two kinds of
brick and two kinds of mortar it would be still more crowded.

330

The chimney, Fig. 81, being now plain, straightaway work
above the set-off from octagonal to round, the outside scaffold
is no longer used.

331

A large mercury plumb bob must be used each day
after quitting time to test the accuracy of the bricklayers' work. The bob must be supported by the smallest of
piano wire. It must be small, so that it will afford the least
surface to the upward draught of air that tends to sway it. It
must be made of wire so that the plumb bob will not dance
or untwist, as it would with a string.

332

At this point, Fig. 82, the chimney's daily growth aver-
aged about 6 ft. in height. The number of masons that can be
used to best advantage on a chimney depends on the number
of pilaster guides to the core, as well as on the amount of
room inside.

333

The plan of the chimney should determine the
number of men to be placed in the chimney. When determining this number, remember that not less than one or more
than two tenders will be required to lift the stock off the
elevator.

334

Give the working foreman the custody of the batter sticks. These should have marked on them at what heights each is to be used. These sticks should be 3 ft. 6 ins. long and 1 in. wide. They should vary in thickness from ⅛ in. at one end to ⅛ in. + the batter in 3 ft. 6 ins. This batter stick is to be attached to the mason's plumb rule with three small screws.

335

Note that the smoke flue is on the opposite side from the

Fig. 82.—Average Daily Growth, Six Feet.

temporary hole left for carrying in stock. Note that the clean-out door opening is under the smoke flue. Locate the temporary opening exactly as shown here.

336

Note the "peach basket," Fig. 83, resting on the brick collar under the head of the chimney. The peach basket will save more than enough of the bricklayer's time to pay for its cost, and the chimney head will be absolutely true in shape as a result of its use. See also Fig. 17.

Fig. 83.—"Peach Basket" Used as a Template for Constructing the Head.

337

Get permission to build a collar under the head of the chimney, to support the "peach basket."

338

Do not permit the makers of the iron cap to ship it until some representative of our firm has seen the complete cap set up and all bolted together.

339

Each piece or section of the iron cap must be self supporting on the wall. If it is not shown thus on the plans, notify the office immediately.

340

Have the cap drilled and tapped with standard thread before leaving the shop, to support the attaching device for the lightning rod.

CHAPTER X.
MORTAR.

341

A very economical method for digging sand and loading carts is shown in Fig. 84.

342

The carts are driven into a trench-like depression, over

Fig. 84.—An Economical Method of Loading Sand Into Carts.

which is a bridge. Across this bridge two-horse drag scrapers are drawn.

343

These empty the sand through a hole in the bridge into the carts underneath.

344

When screening sand, use a netting of long vertical spacing and narrow horizontal spacing, as it will screen many times faster than old-fashioned screens of approximately square openings.

345

The vertical wires should be of tempered steel, so that they will not get out of line.

346

Keep the sand screens in perfect repair, or large pebbles will get into the mortar and delay the bricklayers. It is much cheaper to buy new sand screens than to use damaged sand screens that will cause bricklayers to remove pebbles from the mortar on the wall one at a time with a trowel.

347

The right amount of sand to put into mortar is a question that has interested the leading authorities on this subject throughout the world for years. If more than the right amount necessary to enable the bricklayer to work at his fastest speed is used, the bricklayer will drop more mortar on the ground and waste more time in attempting to bed the brick and to butter the end joints with a mortar that is too sandy to work well, than the value of the lime or cement saved. The workmanship will obviously be inferior to that obtained with mortar that will work easily.

348

Again, if the mortar has too little sand, it becomes sticky, retards the mason's speed, and has more cement or lime than is necessary to properly coat all of the surfaces of the particles of sand and to fill the voids between them.

349

It is injurious to the quality of the mortar to use more lime or cement than just enough to fill the voids and to coat thoroughly each and every grain of sand. The more material is used that is not in its final condition at the time it is used, the more swelling or shrinking, or deterioration and efflorescence will result.

350

These facts must be taken into consideration at the time the voids in the sand are measured and mortar prisms are tested to determine the best and most economical proportioning.

351

Lime mortar must be kept wet while slaking. In addition to water hose, supply a water barrel, so that water can be dashed by the pailful upon any lime that is crumbling or burning. Lime loses strength if allowed to burn.

352

The making of lime mortar must commence early enough to have it at least two weeks old before using. This rule must be followed notwithstanding the fact that the making up of considerable quantities of mortar ahead of time is expensive, because of the extra handling, and the greater labor of tempering up. The older the lime mortar the better the work.

353

Lime mortar must be tempered until all the white spots in it disappear. Otherwise these spots would swell and break the initial set of the mortar after the bricks are laid.

354

Cement must not be added to slaking lime mortar. Cement must be thoroughly mixed dry with sand before it is added to lime mortar, and just before it is to be used.

355

Cement mortar must be used as soon as mixed, unless fat mortar is more desired than strength.

356

The theory most widely accepted among cement experts is that cement, in setting, forms microscopic interlocking crystals. These crystals, if broken while forming, will never properly reunite. Therefore, cement, whether in mortar or concrete, should never be disturbed after it has once begun to set.

357

Give mortar men, or other men most faithful to our interests, first chance on all overtime work, tempering mortar, etc.

358

A mortar bed should always be so located that the sand can be hauled up to and dumped near the long side of the bed. The mortar must be shoveled out of the mortar bed on the side next to the building, if the mortar bed is located outside of the building. If the mortar bed is inside the building, the mortar should be shoveled toward the elevator.

Fig. 85.—Correct Layout for Two Gangs of Mortar Makers.

359

On jobs of considerable size it is always advisable to build a light roof over the mortar beds, Fig. 85. The mortar makers will be able to do more work if they are protected from the rain and sun.

360

Wherever possible, several mortar beds should be arranged near one another, or at least under the same conditions, so that the same number of men can be put on each bed. The

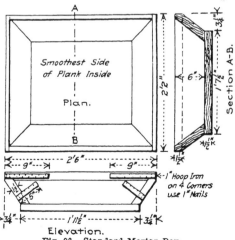

Fig. 86.—Standard Mortar Box.

foreman can then have a continuous athletic contest on mortar making.

361

The mortar box should be well made in accordance with Fig. 86. When two kinds of mortar are used on the same scaffold, a temporary partition in the mortar box will give the most economical results, Fig. 87.

362

Further rules for the handling and placing of the mortar box will be found under Scaffolds and the Gilbreth Scaffold.

Fig. 87.—Gilbreth Scaffold on Overhand Work with Two Kinds of Mortar and Two Kinds of Brick.

363

As bricklayers' wages are so high compared with the cost of mortar, it is always economical to use outside mortar for filling in the wall, rather than to have the bricklayer stand idle while waiting for common mortar.

364

When using cement mortar made of cement and sand and no lime, the bricklayers will do more and better work if a tender is kept on the stock platform tempering the mortar to just the right consistency for the bricklayers. He will also be

very useful keeping boxes filled with mortar and dividing up
mortar between boxes at quitting time, so that all mortar will
be used up. See Fig. 88.

365

Provide him with a short-handled hoe, shovel and pail.
Have him throw away any mortar set too hard to temper with
a shovel or hoe.

366

A small proportion of lime putty, say not less than a pailful to one barrel of cement, will improve the quality of any

Fig. 88.—Tender on Stock Platform to Temper the Mortar.

brick wall, as it insures more than enough better workmanship to compensate for any decrease in the strength of the
mortar. It will also delay the set of the cement enough to
insure the mortar being used up prior to the first set.

367

It is not enough to order lots of square-pointed and
round-pointed shovels. Shovels must be ordered of the size
and shape that are best adapted for the different classes of materials to be shoveled.

Fig. 89.—Rust Spots on a Structural Steel Column.

368

Square shovels must be used for mortar. A tender can shovel sands and all kinds of mortar faster, and can temper up mortar faster with a square shovel than he can with a round-pointed shovel.

369

Figs. 89 and 90 show bad rust spots on structural steel columns that were buried 14 years, from 1892 to 1906, in the exterior brick walls of the Mutual Life Insurance Co.'s eight-story office building at San Francisco, Cal. This building was practically uninjured by the earthquake of April 18, 1906, but was rendered unfit for use by the fire that followed.

370

We removed the upper six stories of this building, and we noted carefully the condition of all of the steel frame.

371

The steel frame was in perfect condition, so far as rust was concerned, except in a few places, probably not more than a dozen places, which were like those shown by Figs. 89 and 90.

372

All of the rust spots occurred where the mortar was not filled closely around the frame. There were no rust spots whatever on any place that was covered with mortar, except in a few minor cases, where the rust apparently had not been completely scraped off before painting. Even in these places the rust seemed to be much retarded by close contact with the mortar.

373

As this steel frame is one of the first used in a high building in the west, and has, consequently, been buried longer than any other that has been carefully inspected when taken down, the results are important to those interested in brickwork.

374

The following rules are the result of our knowledge gained on this building:

(a) All steel, whether painted or not, must be buried in mortar.

(b) All kinds of mortar tend to help to preserve steel from rusting, but Portland cement and sand are undoubtedly the best for this purpose.

(c) All structural steel and iron must be completely

Fig. 90.—Rust Spots on a Structural Steel Column.

plastered by the bricklayer at least 1 ft. ahead of the brick-work.

(d) All brick must be laid with a shove joint against all steel and iron of every description.

375

We realize that this will add to the expense of brickwork, but it must be done thoroughly, regardless of the expense.

376

If the bricklayer sees any evidence of rust on steel or iron work, whether or not it has been painted over, he is to call the brick foreman's attention to it. He, in turn, shall call the superintendent's attention to it, immediately.

377

Care must be taken when wooden beams are built into brickwork that mortar does not touch the wood, as it is likely to cause dry rot. Any kind of mortar will cause deterioration of wood.

378

Whenever possible build the brickwork so that the air can circulate freely around all buried woodwork.

379

Wash mortar from cut stone before it sets too hard.

380

When bedding long stones that are not loaded evenly, for example, sills under openings, care must be used to bed them under the ends only.

381

No mortar must be put under any place except the ends until the entire wall is built and all shrinking has taken place. Otherwise the stone will be broken in two by the uneven loading.

CHAPTER XI.
BRICKS.

382

When writing out the order at the time of buying bricks, be careful about the local terms that define quality. For instance, the terms "light-hard," "hard," "salmon," "benches," "arches," "up and down," "run of kiln," and similar expressions, often have different meaning in different localities.

383

The only safe way to prevent misunderstandings is to have a few sample loads of brick stored in some place where they can be used for reference, to settle all disputes.

384

When buying brick, always give preference to that brick-yard that can show a long record for successes and has produced brick that have been actually exposed to the weather for many years with no sign of deterioration.

385

Other things being equal, the brick that will soak up the least amount of water should be given the preference when buying brick. Where several kind of common brick are used on the same job, the superintendent must use those brick that are least impervious to water where such brick are the most needed.

386

To determine the absorptiveness of a brick, provide a cylindrical glass vessel barely large enough to permit of holding a brick on end, and about twice as high as the brick is when on end. The vessel must have the same horizontal cross sectional area at all heights. Paste a strip of paper vertically on the outside of the glass. Mark off the paper with horizontal lines to any equal divisions, such as 1/16 in., and number them.

387

Proceed as follows:

(a) Fill the glass tank about half full.

(b) Note the exact height on the paper strip of the water level (such as 40).

(c)   Immerse the brick on end and read the height of the water again (such as, say 60).

(d)   Read the height of the water again after the brick has been immersed 24 hours (say at 55).

(e)   Take out the brick and then read the height of the water again (say at 35).

388

This will show that this brick is equal in size to 20 points of water, and that it will absorb five points of water or 25 per cent of its own bulk of water.

389

By this method any number of brick can be compared.

390

If the water does not recede when the brick is taken out the exact number of points that it raised when the brick was immersed, there has either been evaporation, or else the device has been tampered with.

391

The brick must be perfectly dry to start with, or the test will not be accurate.

392

Tests of absorption must always be made in this way, never by weighing the brick before and after immersion; for, while this latter method has been very extensively used, it is very inaccurate.

393

If any round-nosed, octagon, or other special brick are needed, they must be ordered at the earliest possible moment, to prevent the job from being delayed later, while the special brick are being baked.

394

For arches, etc., it is always economical to buy those brick that can be cut the easiest, regardless of their price per thousand.

395

Oftentimes a few thousand special brick can be bought for work that is to be cut. These will save a large sum over cutting the regular cull that is used on the straight work on the balance of the wall.

396

No bats are ever to be used on our work, except those bats that are made by unloading and handling whole brick on the work.

397

This will provide just about enough bats to piece out as they are needed in filling the interior of walls and piers.

398

See that the bricklayers pick up bats when a piece of a brick is wanted instead of breaking a whole brick to get a bat.

399

When wooden brick are required, be sure that they are about the height of two horizontal joints thicker than the clay brick with which they are to be laid. They must be dovetailed —that is, at least 2 ins. longer on the back than on the front so that they will not pull out.

400

For wooden brick pick out the local wood that shrinks and swells the least with moisture and is least subject to dry rot when laid in contact with mortar.

401

Bricks vary in size in different parts of the country. In this book bricks are spoken of as being approximately 8 ins. long, 4 ins. wide and 2½ ins. thick, not because these are the average measurements of merchantable brick, but because it is easier to handle this size in this bricklaying system.

402

We realize that there are as many brick made about 9 x 4¼ x 2¾ ins., as of any other size. The Roman, or Pompeian, size is 12 x 4 x 1½, and the New England water struck brick of 7½ x 3½ x 2⅛ ins. are also numerous; but so far as this bricklaying system is concerned, the rules apply to one size just as well as to another.

403

A vertical tier of brick is herein called "4 ins." of brick. The "closer" at the jamb is spoken of as the "2-in. piece."

404

A brick with a piece cut off one end is called a "three-quarter brick," although it may be considerably longer or shorter than three-quarters of a brick.

405

Some brick have a depression on the top, some on the bottom, and some on both top and bottom surfaces.

406

If the depression comes on one surface only, make it come on the top surface if that is possible.

407

If there is a depression in the bottom surface of a brick, it must be filled with mortar just before the brick is laid, or the air will surely be pocketed in this depression by the mortar, which will prevent the brick from being properly bedded.

408

The depression is supposed to be of value on account of the doweling effect of the mortar, but, in reality, it is put in by the brickmaker to save clay and weight in transportation. It also permits baking hard to the center of the brick with less fuel.

409

The alleged value of the doweling effect of the mortar is more than offset by the uneven bearing that the brick generally gets in practice.

410

Red face brick are culled for two reasons:

      (A) For size.

      (B) For color.

411

Bricks made in the same yard may be made in the same size molds; but, when they are baked, they shrink and color according to the amount and duration of the heat. Consequently, culling them for color also culls them at the same time for size.

412

Bricks can be culled in the shade much easier and faster than in the sunlight. This should be taken into consideration when selecting the place for unloading the face brick.

413

Face brick should be bought with an agreement that they be delivered face up at the job. This means that they must be handled face up from the time they leave the brick kiln till they arrive at the job.

414

All packed brick should be unloaded with iron brick clamps. See Fig. 91.

415

When the brick are piled on the ground, all in one direction, they must be face up, and so placed that the man who culls them can stand at the ends of the brick and dispose of them without walking, as fast as he takes them out of the pile.

416

The unculled brick shown in Fig. 71 are piled exactly right for fast culling.

417

Piling culled brick in small separate piles near the pile of unculled brick is much cheaper than keeping all of the piles of culled brick separate, as this latter method requires much walking. They will not be mixed up if the piles of different culls have plenty of chalk marks showing which culls they are.

418

The culler should stand at the side of the pile facing the ends of the brick. He should use a regular bricklayer's brush and sweep the dust off the entire length of the top row of bull headers (brick on edge), before culling any brick in that row.

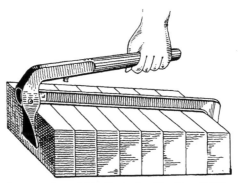

Fig. 91.—Iron Brick Clamp for Handling Brick.

419

Three to five culls are all that are ever expected. The lightest brick are first culls, and the darkest are fifth culls.

420

The culler picks out first that cull of which there are the fewest. If the brick are mostly cull number one, he picks out the fifth cull, and so on.

421

It is obvious that this is the best method of procedure, because, after the few have been picked out, the balance can be handled in bulk from the unculled pile to the various culled piles by low-priced laborers, instead of by the high-priced skilled cullers.

422

Bricks must be put in their individual piles, face up. When the tenders carry the brick to the mason they must deposit them on the mason's stage face up.

423

Bricks should be kept covered until they are culled, as a difference in dampness will deceive the man culling them as to their real color.

424

See that the men who are handling these brick pick them up with both hands simultaneously.

425

All brick must be wet with water thoroughly just before laying, except in frosty weather.

426

On work that is liable to freeze within a week after the brick are laid, the bricks must not be wet, as the freezing of the water in the brick will do more than enough injury to the work to offset the benefits of the wetting of the brick.

427

Where there is no danger of freezing, the brick must be wet almost to the point where they will "run" on the mortar.

428

There are four reasons for wetting the brick:

(a) The bricks will be better bedded on the mortar under them.

(b) They will adhere to the mortar better.

(c) A dry brick soaks water out of the mortar quickly. Mortar will not set properly unless it dries out slowly.

(d) Wetting brick washes the dust from them and clean brick present the best surface for mortar to adhere to.

429

Do not wet brick to a point where they will run out of place when laid in the mortar.

430

If they begin to run do not try to hammer them back to place, as the tapping on the brick tends to temper up the mortar more, and it will run still more.

CHAPTER XII.
BRICKLAYERS' TOOLS, ETC.

431

The tools which an apprentice requires are stated under training of apprentices.

432

Every bricklayer should provide himself with a long and with a short plumb rule. The short plumb rule should be from 1 ft. 4 ins. to 1 ft. 6 ins. long, with one plumb glass and one level glass. The long plumb rule should be not less than 3 ft. 6 ins. long, nor more than 4 ft. long, unless it is to be used with a plumb bob instead of spirit glasses. It should not be over 3¾ ins. wide nor more than 1⅛ ins. thick. It should have a spirit level, with at least two plumb glasses in it, placed at opposite ends and so arranged that at least two can be read at the same time.

Fig. 92.—Brick Jointer.

433

Plumb bob rules are better than an inaccurate spirit glass, but are too slow for general practice. They are difficult to use where the wind blows on the bob.

434

On every job a plumb straight edge shall be maintained, for correcting the mason's levels and plumb rules. Every plumb rule must be tested at least once per month.

435

A jointer, at best, gets dull and loses its shape quickly, due to the wearing qualities of the sand of the mortar. It should, therefore, be made with the hardest temper obtainable. Fig. 92 shows a typical jointer.

436

When gaging and marking 2-in. pieces and three-quarter pieces, a gage similar to A, in Fig 93, saves the most time, as the brick can be marked with a pencil quickest with this kind of a gage.

437

The appearance of any 8-in. wall can be much improved by having all the headers of exactly the same length. Headers of the right length can be most quickly selected by using a wooden gage like B, in Fig. 93.

438

A cutting out hammer should weigh not less than 3¾ and not more than 4 lbs., exclusive of the handle. It should have a hole in the head at least 1 in. long and at least ½ in. wide. The handle should increase in size from the head to the end.

439

There are many other types of hammers and handles that can be used for cutting out but the type above described

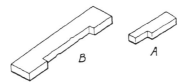

Fig. 93.—Gages for Measuring Brick.

will enable a first-class man to do the most work with the least fatigue. Cast steel hammers are satisfactory and are much cheaper than forged hammers.

440

Every bricklayer is to be loaned free two trowels of the Gilbreth pattern. He is either to return these trowels when he leaves the job, or he is to have their value deducted from his wages. The value of the trowel is to decrease 5 cts. per week for the time he has used it.

441

He is to use the smaller trowel on the exterior 4 ins. of the wall, and he is to use the larger trowel on the interior of the wall.

442

A bricklayer's set is used to cut brick to an exact line.

443

Sets are usually made like Fig. 94, but there is no advantage in having them in this particular shape and they cost

much more than when made like Fig. 95.   In fact the shape
shown in Fig. 95 strikes a better blow and it is easier to control
the direction of the plane of fracture.

444

If the bottom surface is made 60° with the straight side,
the plane of fracture will be about in the plane of the straight
side.

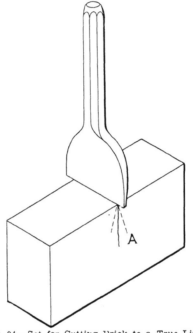

Fig. 94.—Set for Cutting Brick to a True Line.

445

Leaning the set forward slightly will make the fracture
"cut under," as on the dotted line A in Fig. 94, enough to clear
the brick that is laid against the cut brick without further cutting or trimming of the brick after it is cut by the set.

446

The straight side of the set should always be **toward the**
side of the piece of brick to be used, and the bent **face of the**
set toward the piece of brick that is to be wasted.

447

Sets must be kept sharp.   All bricklayers' tools must

be kept constantly in order by a tool sharpener without expense to the bricklayers. There is no more reason for the present custom of bricklayers being obliged to have their own tools sharpened in their own time than there would be reason to make carpenters do likewise.

448

Until comparatively recently, the right temper on ordinary qualities of tool steel for brick tools was obtained best by heating a tool only to a cherry red, then dipping the steel into brine the instant that the last straw color leaves, and when the pigeon blue comes to the end of the cutting edge.

Fig. 95.—Design of a Handy and Low Priced Set.   Fig. 96.—Method of Using a Set.

449

To-day the directions of the makers of the particular kind of steel used must be complied with in order to obtain the best results.

450

If the brick is of a nature that is hard to cut accurately with one blow of a hammer on the set, time can sometimes be saved by cutting off the back corner with the head of the hammer before trying to break off the front corner with the set.

451

When cutting a brick with a set, put the brick in a line in and out with the body, with the piece of brick to be wasted furthest away. See Fig. 96.

452

It is not necessary that the brick be actually broken into two pieces by the set. Furthermore, striking more than one blow on the set is apt to cause the face of the brick to flake up,

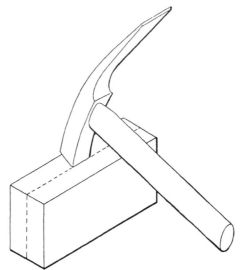

Fig. 97.—Splitting a Brick with the Head of a Hammer.

or injure the face so that it may flake up from the weather after the brick is laid in the wall.

453

One hard sharp rap on the set should be sufficient to develop sufficient weakness in the brick to enable the brick to be broken by one or two sharp flat blows of the hammer on

Fig. 98.—Position of Brick and Hammer When Splitting.

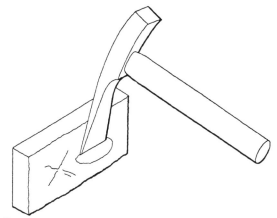

Fig. 99.—Cutting Off Lumps with the Pean of the Hammer.

one of the bed surfaces of the brick in line where the line of fracture is desired.

454

A set will make the straightest line in the shortest time on the cut edge of a brick, but the cut can be made with the head of the hammer (see Figs. 97 and 98). The pean is to be used only for cutting off humps (see Fig. 99) and if the new edge is not exactly straight it can be trued by striking the

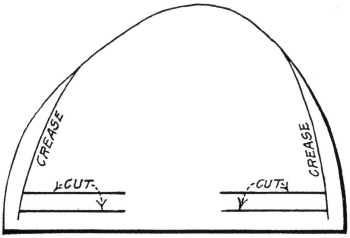

Fig. 100.—Full Sized Pattern for Hand Leather.

projecting points with the flat surface of the face of the hammer.
455

There are various devices for protecting worn fingers and thumbs.
456

Mittens are too clumsy to permit of quick work. Gloves and rubber finger cots make the fingers tender. The best device is a piece of calf skin, if it is cut the right shape. If the leather is cut exactly like the patterns (see Figs. 100 and 101), it can be worn without any hindrance whatever, while if it is cut after the pattern used in a brickyard it is a serious

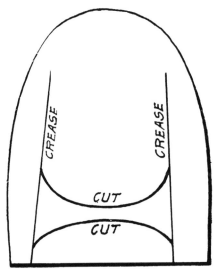

Fig. 101.—Full Sized Pattern for Thumb Leather.

detriment to speed and quality of the work. If the leathers are cut and creased exactly the same size as the patterns, and worn with the bands exactly where shown, with the bands half way between the joints of the first and little fingers (see Fig 102), the leathers will not droop forward and get covered with mortar between the fingers and the leathers.
457

The thumb leather laps over in a way that protects the thumb where the band is cut out.

458

The day has arrived when bricklayers for their best interests must depart from the ancient custom of handling mortar from the mortar box to the wall with a trowel only.

459

New devices must be adopted by this trade to cut down the cost of brick work, or bricklaying will become a lost art. To-day the makers of cement are producing the best cement that the world has ever produced, at a cost of less than 70 cts. per barrel of 380 lbs. at the mill.

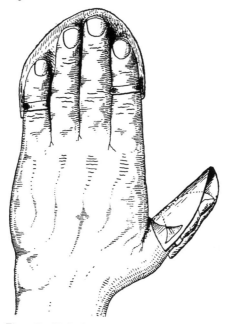

Fig. 102.—Method of Wearing Hand Leathers.

460

The cost of sand for a cubic yard of concrete is but little more than for a cubic yard of brickwork.

461

The cost of broken stone or gravel is much less than the cost of brick of equal bulk.

462

The cost of measuring, feeding, mixing, conveying and placing concrete, together with the cost of the forms, is about

Fig. 103.—Fountain Trowel with Hand Trowel for Handle.

the same per cubic yard as the masons' and tenders' time for cubic yard of brickwork.

463

Furthermore, bulk for bulk, gravel concrete walls without any reinforcement are much stronger and more waterproof, and fully as good non-conductors of heat and cold as the best brick walls. Now that reinforced concrete enters into the con-

Fig. 104.—View Into Trowel Showing Slat for Discharge of Mortar.

struction of nearly every large undertaking, and a certain
amount of concrete plant is therefore necessary, anyway, concrete can be substituted at a considerable saving, in many
places.

464

The time has come when bricklayers must awaken to the
fact that the very existence of their craft is at stake. Means
must be adopted to compete with this oldest yet newest material of construction, concrete.

465

In many places in America, bricklayers have already come

Fig. 105.—Rear View of Fountain Trowel.

to the realization of this fact, and in some states they will lay
no brick on a building the foundation of which is concrete.
In other states they have agreed to lay no brick on a building
the frame of which is reinforced concrete. These agreements
have only spurred the concrete men to new ideas for finishing
buildings of concrete without any brick at all, and the bricklayers have realized that it does not help matters to boycott
concrete.

466

The history of the world repeats itself. In accordance
with the lessons learned from that history, the cost of common brickwork must either be reduced, or bricklaying will become a lost art.

467

We do not recommend trying to solve this problem by reducing the pay of bricklayers or of tenders, for the reason that they are obliged to lose so much time on account of wet and cold weather, nor do we believe that the number of hours should be increased. No one does, who has ever laid brick continuously for eight hours.

Fig. 106.—Teaching Bricklayers to Spread Mortar for Twenty Brick in Five Seconds with the Fountain Trowel.

468

We do believe that the bricklayer must increase his output. He must remove all obstacles that make for reduced output. He must use every device that will lessen the cost of brickwork. The day has come when a bricklayer must adopt other tools than the trowel for handling his mortar. Not a shovel, not a bucket, nor a hod (for good brickwork cannot be done with mortar dumped into the wall), but a fountain trowel, in which the ordinary bricklayer's trowel is used as a temporary handle, can be used for spreading mortar better than mortar can be strung fom a trowel. See Figs. 103, 104, 105 and 106.

469

Fig. 106 shows a demonstration of spreading mortar for twenty bricks in 5 seconds for a header course with a fountain trowel on a 16-inch wall.

470

The fountain trowel makes for good work. It will handle mortar so soft that it could not possibly be handled with an ordinary trowel.

471

This trowel will handle mortar so soft that it will fill even the joints of brick, laid under the brick and brick method. With soft mortar, the work will not dry out so fast, which makes also for better work.

472

An ordinary trowel is used under the back strap (see Fig. 103). When the fountain trowel is empty the hand trow ¹ can be withdrawn instantly, which permits the bricklayer to use the same trowel as a trowel or as the handle to a fountain trowel without any delay in shifting.

473

This is but one of the new methods that must soon be adopted. We shall welcome suggestions that will enable us to produce brickwork at a price low enough to substitute it for concrete.

CHAPTER XIII.

LINES, PLUMBS, AND POLES.

474

There is no one thing that will assist the bricklayer to lay his brick so accurately and quickly to a straight line and plumb face as a very tight line.

475

The old-fashioned English "Linen Mason's Line" should be obtained for this work, no matter where the work is located. This line is the most expensive of all varieties, but will prove economical in every case. Keep your job constantly provided with it.

476

This line is carried in stock in three sizes. Use medium size for common brickwork, and fine for face brickwork.

477

The foreman should see that lines are strung to all plumb corners where there is any chance overhead to attach the lines. This is generally neglected by foremen. Many of the bricklayers think they can do their work as well and as fast with a plumb-rule as with lines. We have never seen one who could, but whether they can or not, lines must be strung.

478

The line should be ⅛ in. out both ways from the brick.

479

After the brick are laid, they are to be tried for straight and for plumb with a plumb-rule, to make sure that the nailing place overhead has not been shifted and that no brick is crowding the line.

480

Provide lines for all pilasters where there is opportunity to attach lines overhead, but inform bricklayers that they must also use their plumb rules.

481

After all lines are in place, sight through from one end to the other, to make sure that they are in line. See Fig. 112A (page 138) and note also the method of staying the metal window frames.

482

In plumbing a corner with a spirit plumb-rule, or with a plumb-bob and rule, it must be held against the brick in a position plumb in itself both ways, that is, also sideways.

483

While there are many ways that a corner can be plumbed, much time can be saved if the plumb-rule is placed in exactly the same place every time. The best place is where the side of the plumb-rule is exactly even, i. e., in the same plane, as the return of the corner. Therefore, on all of our work, the face of the plumb-rule shall be even and flush with the end of the brick on the corner.

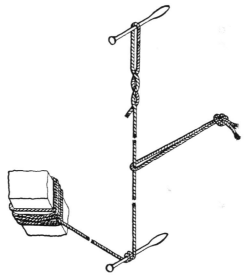

Fig. 107.—Correct Arrangement of the Nails, Line and Trig.

484

Foremen must not permit the men on the lead to build up their leads more than header high. Leads higher than from header to header are unprofitable, and are sometimes the cause of irregularities in the wall.

485

The men on the leads should be told that they are expected to do their bit on the wall with the exception of the header course.

486

While that course is being laid out for them they are to stack up a lead to the next header high.

487

The man on the trig is not expected to carry his trig over two courses above the line. While he must plumb it with a plumb-rule, he must also see that the man on the hauling end of the line sights it for in and out, and also for height.

488

The trig shall be a loop about 6 ins. long, so that the line will haul through it. See Fig. 107.

489

Never make the trig fast to the line.

490

Use a brick on edge to hold the trig in place, and see that the line at the trig is exactly at the top edge of the brick.

491

All brick except trig and lead brick must be laid so that they do not quite touch the line.

Fig. 108.—Correct Method of Splicing the Line.

492

Mason's linen line is too expensive to replace every time it is cut, and even when spliced instead of knotted it is not quite so accurate to lay to and the splice may crowd it off the wall slightly.

493

Men must, therefore, use wire nails and not cut nails to hold the line. These nails should be flattened so as to go into a close vertical joint of the brickwork. See Fig. 107.

494

The nails, splice and trig must be made exactly as shown in Fig. 107.

495

In case the line parts on account of being worn, or by being struck by a trowel, it must be spliced, as shown by Fig. 108, by opening each end in three places and putting each end through the three openings in the other line. It takes more time to do it this way than to tie a knot but it hauls enough more accurately to the line to warrant the time spent.

496

Haul the line to the bottom of all projections and not to the top.

497

The introduction of Portland cement and the recognition of the value of coarse sands has lead to a general tendency toward larger joints in brickwork.

498

The larger the joints the easier it is to secure filled joints and better beds under the brick. This fact should be taken into serious consideration when the foreman is laying out the story pole.

499

Bricks vary much in thickness, even in the same kiln in the same brickyard. Consequently, foremen must constantly watch the work of and confer with the men on the leads, to see if putting in or taking out a course or two of brick on the story pole would assist in making easier, and consequently faster bricklaying on that wall.

500

When the size of a joint is the easiest for the bricklayer, he will bed his brick the most perfectly.

501

The printing on the story pole is to be located in such a manner that it will be easiest read when the pole is right end up. See Fig. 109. This will prevent mistakes due to using the pole wrong end up.

502

It is not always possible to alter the heights of laying up on outside walls, but it is nearly always possible on the inside walls. It is constant attention to these little details that makes the difference between the high pay and the regular pay of brick foremen.

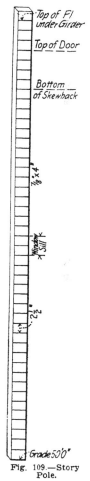

Fig. 109.—Story Pole.

503

All walls should be brought up level in themselves, and at a height about ¼ in. below the beams or plates carrying the floor beams. This will save time in leveling up the floors.

504

Do not permit the man using the engineer's level to use a surveyor's rod for a story pole on buildings, even if he assures you that he can do his work much better with the surveyor's rod.

505

Make all hands depend on the story poles. Their general use will result in many unexpected savings, and danger of mistakes will be eliminated.

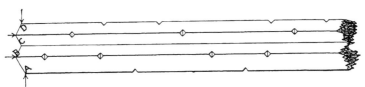

Fig. 110.—Plumb Bond Pole.

506

Notches should be cut in the plumb bond pole as shown in Fig. 110. Marks are made in the last course to correspond with these notches as letters. Bricks laid to the marks insure plumb joints. Figs. 111 and 112 show cross sections of handy shapes for plumb bond poles.

507

All lettering on the plumb bond pole is to be located in such a manner that it can be easiest read when the pole is in the actual position in which it will be used. This will prevent the pole ever being used upside down or end for end and the mistakes that would arise therefrom.

508

When any of the bonds used here are made with a definite border, such as a diagonal header border, a "border pole" can be used with the regular "plumb bond pole."

509

When the border header occurs in a course that is being marked out, the plumb bond pole can be advanced bodily the width of the border or to the next notch on the border pole.

510

The end of the border pole is to be held exactly at the jamb or other plumb mark on the wall.

511

Unless the border runs diagonally for a considerable distance it will be easier for the bricklayers to use a plumb bond pole without the border pole with a notch for each joint of the border bricks.

512

Carry grade marks up one corner of the building, and at a convenient height, about 6 ins. to 1 ft. above each floor. Make level marks on the wall about 10 ft. apart all around each story.

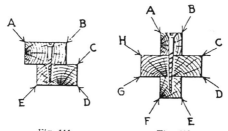

Fig. 111.   Fig. 112.
Suggested Cross Sections of Plumb Bond Poles.

513

Use the level marks for all measuring for heights of all different pieces of construction in this story, and for setting the beams of the floor above; but do not measure from the top of the floor, no matter how accurately it was measured and leveled up.

514

Use the grade marks as described above, for two purposes :

(a) To measure down from, to check up the level of the floor, and thus detect a mistake before it is too late.

(b) To correct errors, even if a part of the building has slightly settled or shrunk. It is a well-known fact that brickwork will shrink as the increased weight is put upon it about $\frac{1}{4}$ in. to the average story height. In fact, some walls will shrink so much that allowance for shrinking must be made in setting beams, one end of which rests on a steel frame and the other end on the brickwork.

515

After a floor has been leveled up, the foreman should sight along the tops of the beams to see if they line up accurately, that is, to see if they are in the same plane.

516

Sight the tops of window frames before they are bricked in.

517

Sight the tops of bearing plates before they are loaded.

518

Sight piers in a row for straightness of the row and for plumb. A great many errors and pieces of bad work can be

Fig. 112A.—Lines Used as Guides for Plumbing Corners.

detected at a time that they can be corrected at small cost, if the foreman and superintendent will make a practice of sighting everything that is in the same plane.

519

There are many cases where bosses have earned a great reputation for "accuracy of eye for plumb and level," when in reality they sighted the bricklayers' lead to the corner of a building across the street when nobody was watching. This sighting and ranging for level and for plumb is good practice for the boss, superintendent and foreman.

520

The best bricklayers, also, when building their leads, sight each course by some level course or object in the distance instead of constantly using a spirit level; but it must here be remembered that two level courses cannot be used for comparison unless they are either exactly level with each other, or else parallel with each other; while any two plumb lines can be used for comparison.

521

All measurements must be made three times.

522

The original measurement must be checked before the bricklayer starts his lead.

523

The second check must be made soon after the bricklayer has actually started the brickwork.

524

The purpose of the second check is to make sure that the bricklayer has actually used correctly the measurements that have been given to him.

525

Wherever there is a definite measure that must be maintained, a templet, or stick, must be made and used to check up the other measurements. For example, in a power station, the exact distance between the walls carrying the crane girder must be checked as built by a wooden pole of the exact length. This pole must be used several times in the length and also in the height of the wall so that, regardless of what other means of measuring have been taken, the pole will check it.

CHAPTER XIV.
MOTION STUDY.

526

The motion study in this book is but the beginning of an era of motion study, that will eventually affect all of our methods of teaching trades. It will cut down production costs and increase the efficiency and wages of the workman. It will, we hope, eventually help to handle the industrial problems which are now being solved by Mr. Frederick W. Taylor, Ex-President American Society Mechanical Engineers. by means of elementary time study, the task and the differential piece rate. (See paper 1003, Trans. Am. Soc. M. E., by F. W. Taylor.) All members of our organization should study carefully the work that is now being done by Mr. Taylor and his collaborators, Messrs. Sanford E. Thompson, H. L. Gantt, C. J. Barth, H. K. Hathaway, and also the cost analyses investigations that are now being made by Messrs. H. P. Gillette and R. T. Dana.

527

There is a tremendous field, in all branches of all mechanical trades, for descriptions and illustrations in print of the best methods used by the best mechanics in working at their trade. We particularly request photographs showing such methods to the best advantage.

528

To be pre-eminently successful: (a) A mechanic must know his trade; (b) he must be quick motioned; and (c) he must use the fewest possible motions to accomplish the desired result.

529

It is a fact beyond dispute that the fastest bricklayers, and generally the best bricklayers, are those who use the fewest motions, and not those who are naturally the quickest motioned.

530

A bricklayer can do no better service for his craft than

to devise methods for laying brick with fewer motions than are at present practiced by bricklayers.

531

We present herein charts of the methods used by our best bricklayers.

532

It would not be feasible to illustrate all the different methods used by all of our bricklayers; furthermore it would be confusing.

533

The purpose of the illustrations is to teach apprentices that a brick can be laid with very few motions, if each motion is made for a certain desired effect, and that a combination of these motions gives the certain desired result.

534

It is a recognized fact among bricklayers, that they use one set of motions when they are trying to exceed the speed of a fellow workman, and another set when they are not especially rushed.

535

When a bricklayer shows an apprentice how to lay brick he invariably teaches the slow method. The result is, the apprentice learns to place the brick in the right place with the right amount of mortar under and against it, but the method used involves a great many more motions than are necessary.

536

The apprentice, after becoming an expert in this way, must then attempt to get out of the slow habits, due to unnecessary motions, and to learn to lay brick by a method that will enable him to complete his portion in the time that is allotted to journeymen.

537

These illustrations will enable the apprentice to earn his money from the first week he starts to work.

538

The rules will narrow down his first lessons to a few vital principles and motions. They show what he should learn first, as well as how he should learn it.

539

These rules and charts will enable the apprentice to earn large wages immediately, because he has here a series of instructions that show each and every motion in the proper sequence. They eliminate the "wrong way," all experimenting, and the incompetent teacher.

540

We do not want any bricklayer not well acquainted with the method and motions herein laid down to waste either his own time or the time of the apprentice teaching the latter.

541

Now as to the journeyman bricklayer, himself, we have a difficult problem to handle. We have found that some bricklayers with good intentions cannot be made to leave off their old habits of making a dozen or more motions per brick, because they have been laying brick in that way for many years. Yet, by hard and continued work, with little time spent in resting, they are able to do a profitable amount of work per day.

542

It is not wise to interfere with this type of man.

543

Again, there is the bricklayer who can adopt any method, but who cannot get such good results from new methods.

544

We must have the best work in spite of all other considerations. Therefore, it is not wise to have him change from the method under which he is most skillful.

545

Another type, which is the commonest of all, is the man who, unconsciously, uses our method when he is rushed, but who, unconsciously, uses other methods when he is not rushed.

546

It is our intention to increase the wages of those men who lay brick in the manner described in this system, because we know that with the usual amount of effort and the same number of motions our method will increase the number of brick laid by two or three times the number laid under unsystematic methods.

547

We shall, therefore, continue to rate our bricklayers by classes, as follows:

(a)   Those who adapt themselves to this system.   Men of this class shall receive a substantial increase above the minimum rate of pay.

(b)   Those who can adapt themselves in part to this system. They will receive more money than the minimum rate.

(c)   Those who are not able to adapt themselves to this system, but who can, by great and constant effort, accomplish a fair day's work. They shall receive the minimum rate.

(d)   Those who do not ever attempt to lay brick in accordance with this system. They shall be employed only when regular bricklayers are scarce.

548

To save all the time possible, and to do the work with the least manual effort, is the purpose of the charts.

549

Apprentices must be taught to make up charts representing their own motions.

550

They must be permitted to use a reasonable amount of time in charting the times of the operation of our best bricklayers, that they may fully compare the bricklayers' methods with the charts in this book, and that they may also see their own shortcomings, by comparison.

Fig. 113.—Exterior Face Tier. Working Right to Left, Spreading Mortar.   Fig. 114.—Exterior Face Tier. Working Right to Left, Cutting Off Mortar Before Brick Is Laid.

551

Foremen must be careful to insist that the rules here given are followed by our apprentices. They will not only lay more brick by following them, but they will also become more valuable additions to our organization. They will make better foremen bricklayers for us than men with a much wider experience who have not been carefully trained under our system.

552 The plan of walls and the motion study charts inserted below cover the twelve following cases of bricklaying:

Plan I. Case 1—Pick and dip, overhand, right to left. For actual motion studies, see Figs. 113, 114, 115 and 116.

Plan I. Case 2—Pick and dip, overhand, left to right. See Fig. 117.

Fig. 115.—Exterior Face Tier. Working Right to Left. Buttering the End of the Laid Brick.

Fig. 116.—Exterior Face Tier, Working Right to Left. Cutting Off Mortar After the Brick Is Laid.

Plan I. Case 3—Pick and dip, inside, right to left, illustrated in Figs. 118, 119, 120 and 121.

Plan I. Case 4—Pick and dip, inside, left to right. See Figs. 6 and 122.

Fig. 117.—Exterior Face Tier. Working Left to Right. Buttering the End of the Laid Brick.

Fig. 118.—Interior Face Tier, Working Right to Left, Throwing Mortar.

Plan II. Case 5—Pick and dip, middle of wall, right to left.

Plan II. Case 6—Pick and dip, middle of wall, left to right.

Fig. 119.—Interior Face Tier, Working Right to Left, Spreading Mortar.

Fig. 120.—Interior Face Tier, Working Right to Left, Cutting Off Mortar Before Brick Is Laid.

Fig. 121.—Interior Face Tier, Working Right to Left, Tapping Down Brick.

Fig. 122.—Interior Face Tier, Working Left to Right, Cutting Off Mortar After Brick Is Laid.

Fig. 123.—Exterior Face Tier, Working Right to Left, Spreading Mortar.

Fig. 124.—Exterior Face Tier, Working Left to Right, Cutting Off Mortar After the Brick Is Laid.

Fig. 125.—Exterior Face Tier, Working Left to Right, Cutting Off Mortar After the Brick Is Laid.

Fig. 126.—Exterior Face Tier, Working Left to Right, Buttering the End of the Laid Brick.

Fig. 127.—Interior Face Tier, Working Right to Left. Spreading Mortar.

Fig. 128.—Interior Face Tier, Working Left to Right, Spreading Mortar.

Plan III Case 7—Stringing mortar, overhand, right to left. See Fig. 123.

Plan III. Case 8—Stringing mortar, overhand, left to right. See Figs. 124, 125 and 126.

Plan III. Case 9—Stringing mortar, inside, right to left. See Figs. 127 and 128.

Plan III. Case 10—Stringing mortar, inside, left to right. See Figs. 129 and 130.

Fig. 129.—Interior Face Tier, Working Right to Left, Throwing Mortar.

Fig. 130.—Interior Face Tier, Working Left to Right, Spreading Mortar.

Plan IV. Case 11—Stringing mortar, middle of wall, right to left, with shove joints.

Plan IV. Case 12—Stringing mortar, middle of wall, right to left, with brick and brick method.

553

Chart No. 1 shows the number of motions required in the process of laying a face brick under the old fashioned method provided no two motions are done simultaneously.

While chart No. 1 shows about all the defects of all brick-layers, it is not probable that any one first class bricklayer would use all of the eighteen operations as shown on this chart.

554

Cases 1 to 4.—See Chart I. for explanation.

555

Case 5. Traveling Right to Left.—Pick the brick nearest at hand, and carry the brick and mortar in a straight line from the stock platform to place.

556

Deposit the brick in the furthest unfilled 4 ins., or tier, and drop the mortar in the next furthest 4 ins., or tier, at exactly the same time.

557

The momentum of the brick will help to shove the joint full of mortar.

558

On the work in the middle of the wall, never cut off the mortar that is pushed up above the top of the bricks oftener than every ten bricks.

559

Deposit it in the middle of the wall. It takes no more motions to cut off the mortar from the top of ten bricks than from the top of one brick.

560

Case 6. Traveling Left to Right.—The brick and the mortar should not only be picked up at the same time, but they must also be deposited at the same time. The old practice of dropping first the mortar and then the brick must never be used on our work.

561

If the brick and mortar are deposited in the same tier, they can be carried in almost a straight line from stock plat-form to place, and can be deposited at exactly the same time.

562

If the brick is carried fast from the stock pile to place, the momentum will help to shove the joint full of mortar.

Operation No.	The Wrong Way. Motions per Brick.	The Right Way. Motions per Brick.	PICK AND DIP METHOD. THE EXTERIOR 4 INCHES (LAYING TO THE LINE).
1	Step for Mortar.	Omit.	On the scaffold the inside edge of mortar box should be plumb with inside edge of stock platform. On floor the inside edge of mortar box should be 21 in. from wall. Mortar boxes never over 4 ft. apart.
2	Reaching for Mortar.	¼	Do not bend any more than absolutely necessary to reach mortar with a straight arm.
3	Working up Mortar.	Omit.	Provide mortar of right consistency. Examine sand screen and keep in repair so that no pebbles can get through. Keep tender on scaffold to temper up and keep mortar worked up right.
4	Step for Brick.	Omit.	If tubs are kept 4 ft. apart, no stepping for brick will be necessary on scaffold. On floor keep brick in a pile not nearer than 1 ft. nor more than 4 ft. 6 ins. from wall.
5	Reach for Brick.	Included in 2.	Brick must be reached for at the same time that the mortar is reached for, and picked up at exactly the same time the mortar is picked up. If it is not picked up at the same time, allowance must be made for operation.
6	Pick up Right Brick.	Omit.	Train the leader of the tenders to vary the kind of brick used as much as possible to suit the conditions; that is, to bring the best brick when the men are working on the line.
7	Mortar, Box to Wall.	¼	Carry stock from the staging to the wall in the straightest possible line and with an even speed, without pause or hitch. It is important to move the stock with an even speed and not by quick jerks.
8	Brick, Pile to Wall.	Included in 7.	Brick must be carried from pile to wall at exactly same time as the mortar is carried to the wall, without pause or jerk.
9	Deposit Mortar on Wall.	Included in 7.	If a pause is made, this space must be filled out. If no pause is made it is included in No. 7.
10	Spreading Mortar.	Omit.	The mortar must be thrown so as to require no additional spreading and so that the mortar runs up on the end of the previous brick laid, or else the next two spaces must be filled out.
11	Cutting off Mortar.	Omit.	If the mortar is thrown from the trowel properly no spreading and no cutting is necessary.
12	Disposing of Mortar.	Omit.	If mortar is not cut off, this space is not filled out. If mortar is cut off keep it on trowel and carry back on trowel to box, or else butter on end of brick. Do not throw it on mortar box.
13	Laying Brick on Mortar.	¼	Fill out this space if brick is held still while mortar is thrown on wall. When brick is laid on mortar it presses mortar out of joints; cut this off only at every second brick.
14	Cutting off Mortar.	Every 2nd brk. ½	It takes no longer to cut mortar off two bricks than one,
15	Disposing of Mortar.	Butter ¼ End Joint.	When this mortar is cut off it can be used to butter that end of the last previous brick laid or it can be carried on the trowel back to the box.
16	Tapping Down Brick.	Omit.	If the mortar is the right consistency, with no lumps in it, and the right amount is used, the bricks are wet as possible without having them run, no tapping with the trowel will be necessary.
17	Cutting off Mortar.	Omit.	If the brick must be tapped, hit it once hard enough to hammer it down where it belongs. Do not hit the brick several light taps when one hard tap will do.
18	Disposing of Mortar.	Omit.	Do not cut off the mortar oftener than every second brick, and when you do cut it off do not let it fall to the ground; save it; keep it on the trowel, and do not make another motion by throwing it at the box. Carrying it to the box does not count another motion.
	18	4½	Total number of motions per brick.

Chart 1.—Pick and Dip Method. Laying to the Line.

PICK AND DIP METHOD. CENTER OF THE WALL.

Operation No.	The Wrong Way. Name	The Wrong Way. Motions per Brick.	The Right Way. Motions per Brick.	PICK AND DIP METHOD. CENTER OF THE WALL.
1	Step for Mortar.	1	Omit.	If a step is necessary the mortar boxes are too far apart.
2	Reaching for Mortar.	1	1	When reaching for brick and mortar always pick the stock first which is the nearest to the wall. Use those brick that are the farthest away from the wall as a reserve pile to be used only when pile gets small.
3	Working up Mortar.	1	Omit.	
4	Step for Brick.	1	Omit.	If mortar boxes are not more than 4 ft. apart and the bricks are all piled near the wall no step will be necessary.
5	Reach for Brick.	1	Included in 2	
6	Pick up Right Brick.	1	Omit.	The only selection of brick for work in the middle of the wall is to pick up those brick that are the least fit for the exterior 4 ins., i.e., chipped, broken, misshapen and discolred brick.
7	Mortar, Box to Wall.	1	1	When conveying mortar from box to wall carry it in the shortest and straightest line possible. Do not pause in the path. Keep the mortar going at an even speed from box to wall.
8	Brick, Pile to Wall.	1	Included in 7.	Convey brick from pile to wall in shortest line, and use momentum of brick to help shove the brick into the mortar.
9	Deposit Mortar on Wall.	1	Included in 7.	Deposit the mortar to the right of the place where the brick is to be laid. The depositing of the mortar and brick can then be done simultaneously without delaying the denositing of the brick.
10	Spreading Mortar.	1	Omit.	The mortar must be thrown so as to require the least amount of effort, to shove the joints full of mortar when the next brick is laid on it.
11	Cutting off Mortar.	1	Omit.	No spreading should ever be required of mortar in the middle of the wall pick and dip method.
12	Disposing of Mortar.	1	Omit.	
13	Laying Brick on Mortar.	1	Included in 7.	Do not lay the brick on the mortar conveyed to the wall at the same operation. Lay the brick in a place to the left of the place where the mortar is denosited and on ton of the previous trowelful.
14	Cutting off Mortar.	1	Omit.	The brick should be shoved only far enough to just bring the mortar to the top of the brick, and no more, and then no cutting off mortar will be necessary.
15	Disposing of Mortar.	1	Omit.	
16	Tapping down Brick.	1	Omit.	
17	Cutting off Mortar.	1	Omit.	
18	Disposing of Mortar.	1	Omit.	
	Total Number of Motions.	18	2	

Chart 2.—Pick and Dip Method, Laying in the Interior The . . .

Operation No.	The Wrong Way. Motions per Brick. ¼ ½ ¾ 1	The Right Way. Motions per Brick. ¼ ½ ¾ 1	STRINGING MORTAR METHOD. THE EXTERIOR 4-INCHES (LAYING TO THE LINE.)
1	Step for Mortar.	Omit.	
2	Reaching for Mortar.	¼	As a large trowel holds mortar enough for four brick, ¼ of a motion is the right amount to allow for one brick.
3	Working up Mortar.	Omit.	Have a laborer keep the mortar at the right consistency by tempering.
4	Step for Brick.	Omit.	If the mortar boxes are not over 4 ft. apart no stepping is necessary.
5	Reach for Brick.	¼	
6	Pick up Right Brick.	Omit.	
7	Mortar, Box to Wall.	¼	Conveying mortar for four brick, equals ¼ motion per brick.
8	Brick, Pile to Wall.	½	Brick in each hand = ½ motion per brick.
9	Deposit Mortar on Wall.	¼	Depositing mortar for four brick at once = ¼ motion per brick.
10	Spreading Mortar.	¼	Spreading mortar for four brick per motion = ¼ motion per brick.
11	Cutting off Mortar.	Omit.	Do not cut off any mortar until brick is deposited on mortar.
12	Disposing of Mortar.	Omit.	
13	Laying Brick on Mortar.	Included in 5.	
14	Cutting off Mortar.	¼	
15	Disposing of Mortar.	¼	Butter the end of last brick laid.
16	Tapping Down Brick.	Omit.	
17	Cutting off Mortar.	Omit.	
18	Disposing of Mortar.	Omit.	
	18	4½	Total Number of Motions per Brick.

Chart 3.—Stringing Mortar Method. Laying to the Line.

Operation No.	The Wrong Way. Motions per Brick.	The Right Way. Motions per Brick.	Stringing Mortar Method. The Center of the Wall.
1	Step for Mortar.	Omit.	
2	Reaching for Mortar.	$\frac{1}{4}$	Mortar enough for four brick at a time = $\frac{1}{4}$ motion per brick.
3	Working up Mortar.	Omit.	
4	Step for Brick.	Omit.	
5	Reach for Brick.	$\frac{1}{2}$	One brick in each hand = $\frac{1}{2}$ motion per brick.
6	Pick up Right Brick.	Omit.	
7	Mortar, Box to Wall.	$\frac{1}{4}$	Mortar for four bricks at once = $\frac{1}{4}$ motion per brick.
8	Brick Pile to Wall.	$\frac{1}{2}$	One brick in each hand = $\frac{1}{2}$ motion per brick.
9	Deposit Mortar on Wall.	Included in 7.	
10	Spreading Mortar.	$\frac{1}{4}$	Spreading mortar for four brick to the motion = $\frac{1}{4}$ motion per brick.
11	Cutting off Mortar.	Omit.	
12	Disposing of Mortar.	Omit.	
13	Laying Brick on Mortar.	Included in 8.	
14	Cutting off Mortar.	Omit.	If the brick is not shoved too far on the bed of mortar the mortar will be shoved just to the top edge of the brick, and no cutting of mortar above top of brick will be necessary.
15	Disposing of Mortar.	Omit.	
16	Tapping Down Brick.	Omit.	
17	Cutting off Mortar.	Omit.	
18	Disposing of Mortar.	Omit.	
	18	$1\frac{3}{4}$	

Chart 4.—Stringing Mortar Method, Laying in the Interior Tiers.

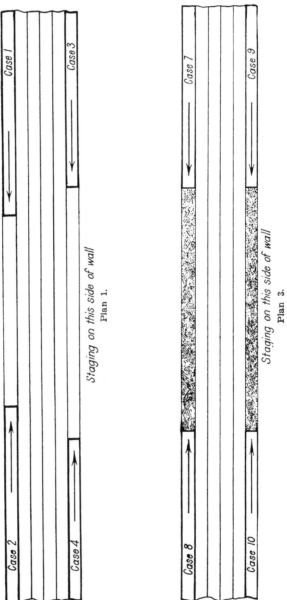

Case 1

Case 3

Case 2

Case 4

Staging on this side of wall
Plan 1.

Case 7

Case 9

Case 8

Case 10

Staging on this side of wall
Plan 3.

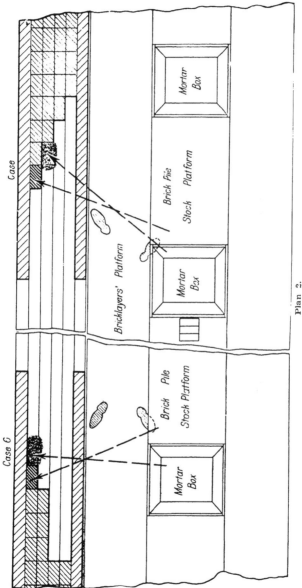

Plan 2.

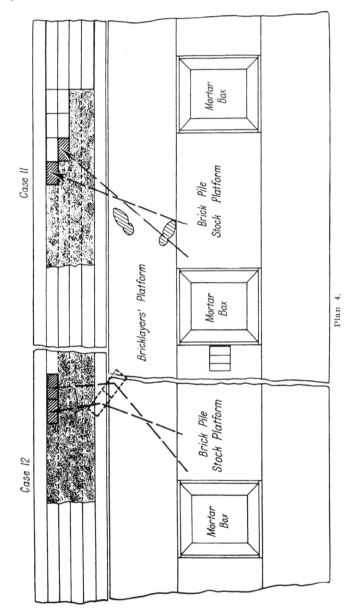

Plan 4.

563

Cases 7 to 10. See Chart III. for explanation.

564

Case 11. Traveling Right to Left.—Pick up the two nearest brick on the scaffold, and carry them to place in as nearly a straight line as possible. Deposit them at exactly the same time in different tiers.

565

Case 12. Traveling Right to Left.—On work where unfilled vertical joints in the middle of the wall are desirable and on work that is to be grouted, pick up the two nearest brick, carry them toward place, and when they are about a foot from place put them together in the air. Then deposit them in the same tier. This makes the air space between them as small as possible, and saves time.

566

You cannot deposit them at the same instant and get shove joints full of mortar, unless they are in different tiers.

567

The following rules are made on the supposition that the work is being done from the Gilbreth scaffolds. When it is done on the floor, or from other kinds of scaffolds, the work should be done as nearly as is possible in accordance with these rules.

568

Stepping for mortar (1) and reaching for mortar (2) must be done exactly at the same time.

569

The same is true of stepping for brick (4) and reaching for brick (5).

570

The apprentice must make it a point to stand where he can pick up his stock with both hands at the same instant with the least effort.

571

After he has found that spot on the bricklayers' platform where each foot should be, he must stand there without stepping, and lay as far as possible in each direction, without making a step or lifting either shoe completely off the platform.

572

We find, after many years of actual practice, that if the scaffold horses are set up anywhere from 10 ft.

to 10 ft. 6 ins. apart, and if two mortar boxes, each 2 ft. 2 ins. wide, are placed in each bay, one of them being hard against the left hand upright and the other divided evenly between that box and the right hand upright, then the bricklayer can stand on any part of the bricklayers' platform and reach into a mortar box and also reach the brick pile.

573

If the horses and mortar boxes are spaced as stated above, operations numbers 1 and 4 can be omitted.

574

We have found that 1 ft. 5 ins. is the shortest distance that the bricklayer can work in comfortably. Consequently, we have designed the patent horse so that when the foot is butted against the wall, the edge of the stock platform is exactly 1 ft. 5 ins. away from the wall.

575

We have also found that the bricklayer picks up his stock with the least fatigue from a platform 2 ft. above the level on which he stands. The same is true of the height of the wall on which he lays the brick. We have consequently made the stock platform 2 ft. higher than the bricklayers' platform. We have arranged the lifting jacks to work on 8-in. notches, so that the stock platform and the top of the wall will be at the same level. This is the most convenient and comfortable arrangement for the bricklayer. It cuts down the distance for reaching for mortar (2), reaching for brick (5), conveying the brick from the staging to the wall (8), and conveying the mortar from the staging to the wall (7).

576

The bricklayer should always pick up those brick first that are on the side of the stock platform that is nearest the wall.

577

He should pick up the mortar from that part of the box that is nearest the wall, in order to reduce the conveying distance.

578

He should use the stock that is far away only when he has none near the wall.

579

Working up the mortar with the trowel (3) should be dispensed with by having a tender on the stock platform with

a water bucket and hoe (see Fig. 88) to keep the mortar at the
right consistency for the speediest bricklaying.

580

Even with a small number of masons, it pays to put a
tender on the stock platform. He can not only temper up
the mortar, but he can devote any spare time to piling up the
brick on the inside of the stock platform with their faces up,
so that the time of picking out the right brick (6), can be
reduced to almost nothing.

581

The time needed to convey the mortar from the staging
to the wall (7), depends not only on the distance that the

Fig. 131.—Spreading Mortar with One Motion.

mortar box is from the wall, but also on the amount of mortar
that is taken at each trowelful. This is one of the points where
the stringing mortar method outclasses the pick and dip
method for speed.

582

Spreading mortar (10) should never require over one motion per brick. (See Figs. 113, 119 and 131.) In most cases
of common brick work, the mortar should be thrown from the
trowel so as to require no further spreading. See Figs. 6 and 7.

583

The apprentice who spreads his mortar with two or
more motions, or strokes of the trowel, should be
watched carefully. He should be made to count the motions that he makes in laying each brick, until he is able to
lay brick in the manner and with the same number of motions
laid down in this system.

584

Cutting off the mortar that projects over the edge of the wall before the brick is laid (11), is entirely unnecessary. (See Figs. 114 and 120.) This should never be done on common brick work.

585

Cutting off mortar after the brick is rubbed into the mortar (14) will furnish the mortar for the next end joint. (See Figs. 116, 122, 124 and 125.)

586

Buttering the end joint (12) should be omitted at process (11). and should be done only at process (14). (See Figs. 115, 117, 126 and 132.)

Fig. 132.—Buttering the End of the Brick in Hand.

587

Rubbing the brick into the mortar (13) will require almost no time or effort, certainly not over one-half a second, provided the joints are the right thickness and the brick are properly wet. (See Figs. 116 and 122.)

588

Foremen must, therefore, see that the brick are constantly kept at the right degree of wetness, not only to insure good work but also for speed.

589

The foreman must personally lay out the story pole for those heights of laying that will make the most speed as well as the best work.

590

Mortar on the trowel that is cut off from under the brick should be put on the end of the brick previously laid, for filling the end joint. (See Figs. 115, 117 and 126.) It should never be thrown from the trowel back into the mortar box.

591

Many masons have the habit of constantly throwing the mortar back into the box. A large portion of this daubs up the bricks instead of landing in the box; besides, it means an unnecessary motion every time.

592

Tapping brick down to grade with the trowel should not be necessary, if the mortar is of the right consistency, the brick is wet enough and the joints are the right size; but if a tapping is necessary, tap the brick one hard tap, instead of several light taps. (See Fig. 121.) This reduces the operation of tapping to one motion of not over one-half second.

593

Nearly all bricklayers tap the brick from habit, not because it is necessary.

594

These charts of each case should be used as examples by apprentices as to the methods they should first learn and which motions they should use also as the total of motions for each process.

595

After the apprentice has learned the twelve different processes exactly as shown, he should be permitted to practice any other method that will accomplish the same quality of work in the same amount of time, as there are many different ways of laying brick.

596

In filling in the middle of a wall it is always quicker to lay those brick nearest the overhand side first and those nearest the inside face last. This order will allow the carrying of the brick from the stock platform to the wall with the most uniform speed, without a hitch or a change of direction of the motion.

597

Close watching of bricklayers will disclose the remarkable fact that years of constantly training the left hand to tell by feeling the top side from the bottom side of a brick, forms the habit of turning a brick over in the hand so as to have it right side up, even if it is being laid in the filling tiers. Few bricklayers realize that they do this, as it has become automatic with them to do it for the face tiers.

598

When seen to do this while laying on the filling tiers, they should receive a few reminders that they are not to do so, as it requires just so many more unnecessary motions and fatigues them for no purpose, making them require just so much more rest.

599

Teach them to make absolutely no motions and to have their hands travel no distance that does not give results.

600

In the selection of these methods as adopted here for the training of our young men, we have followed the best of the working methods of the men in our organization—which consist of bricklayers from ·many different nations, who have adapted themselves to the different conditions existing in various parts of the United States.

CHAPTER XV.

METHODS OF LAYING BRICK UNDER SPECIAL CONDITIONS.

601

When laying brick in freezing weather, keep the brick as dry as possible. Heat the brick clear through.

602

Use mortar made hot by the use of hot water and hot sand but do not use hot lime mortar made hot by the slaking of the lime, for hot lime is not completely slaked and if used will continue to slake and disturb the cement after it has commenced to set and decrease or ruin its strength.

603

Use sharp coarse sand. Do not fill the side joints any more than is absolutely necessary, but slush the top course of the wall solid before quitting time, to keep out rain and snow.

Keep the wall backed up solid all the time.

604

Keep the wall covered nights, not only so that the top of the wall will be kept dry, but also so that the water running off the cover of the wall will not drip down on the face of the wall.

605

In very cold weather, the bricks and mortar should be re-heated around a salamander on the scaffold.

606

Special care must be taken to brace work laid in cold weather.

607

If two kinds of brick are on the same wall, it will generally heave toward the side of the thinnest joints.

608

If the sun shines on one side and it is freezing on the other, it will heave toward the sun.

609

The covering must not only be water tight, but must also be made so that the snow will not blow in on top of wall. After every snow storm, tops of walls, sills, projections and scaffolds must be swept off clean, or snow will melt and run into the wall.

610

In cold weather salt may be used in small quantities in the mortar, but it is better to get along without it if possible. It will make more efflorescence on the surface of the face brick when the work dries out, and it keeps the mortar damp for a much longer time than it would be without it.

611

When bricks are laid on a curved vertical surface, the bricks must always be laid with a heavy roll to look right, but there is never any excuse for lipping brick.

612

The correct amount of roll can be determined by putting the lower corners 1-32 in. inside the line of the brick directly under those corners of the ashlar line, and the center of the top edge 1-32 in. outside the ashlar line.

613

If the curve is of so short radius that the two upper corners of the brick project so far as to look badly, then the brick must be shortened until each stretcher is made short enough to reduce the fault or else Flemish bond or Flemish spiral bond must be used.

614

This latter will also reduce the amount that any part of any brick is in or out from the ashlar line.

815

All brick work below ground must be laid without overhanging lips or corbels, or the frost will get a hold on them and heave the brick work.

616

When building a wall up against the wall of an adjoining property, have the face of the wall of the next property surveyed to determine the exact building line of the wall of your building.

617

If the wall of the abutter overhangs or encroaches on our job, notify the owner. If it is clear and sets back from our job, be careful not to encroach. Build the exterior property line face in a manner that will show good work and a good face in case it is ever exposed by the razing of the adjoining wall.

618

If you should follow the lines of the wall of the next property and encroach, the owner of our job may be obliged to cut off or take down his wall when a new building is built from a correct survey on the adjoining lot.

619

If any one orders you to follow any other line than that shown on the drawings, you must receive orders in writing before proceeding with the work.

620

Few bricklayers know how to lay fire brick on surfaces that will be exposed to long continued high temperatures.

621

Engineers who have to construct ovens and furnaces invariably secure the services of a superintendent who has an established record for good work in this line.

622

But it is possible for any bricklayer who will follow directions to do the work. The secret of success is simply to have close joints between the bricks.

623

The brick must be kept in a tub of water, under water, until they are laid in the wall.

624

The mortar, regardless of what it is made, must be as soft as a thick soup. Reground second hand fire brick of good quality that have been taken out of a furnace lining makes the most reliable fire resisting mortar.

625

Do not lay the brick with a trowel.

626

Reach down into the water and pick up a brick.

627

Dip the brick immediately down into the mortar, so that its bottom and one end will be coated with a thin soft layer of mortar.

628

Lay the brick with a hammer instead of a trowel.

629

The main object is to lay the brick with the thinnest possible joint.

630

No mortar at all is much better than a large joint.

631

Never mind how much mortar runs down onto the face of the work; never mind how much mortar runs out of the joints; never mind if the work is not level.

632

Keep the joints as close as possible.

633

The mortar should be so soft, the bricks so wet, and the joints so tight, that five courses of brick should measure no more when laid up in the wall than when stacked up dry.

634

A hot fire must not be built against fire brick that are not perfectly dry, otherwise the brick will be severely injured by the steam in them.

635

The fire brick must have ample time to dry out thoroughly before heating up. If this is not possible it is far better to lay them dry instead of wet.

636

By what is said above it is not meant that a trowel will not be used at all. After a course of brick have been laid with a hammer, there will undoubtedly be some end and side joints that can be slushed full with a trowel, but the trowel should not be used at all during the process of laying the brick.

CHAPTER XVI.

FINISHING, JOINTING AND POINTING.

637

All bricklayers must be told how the wall is going to be finished.

638

Inform the bricklayer whether the wall is to be (a) plastered directly upon its surface, (b) furred, lathed and plastered, (c) painted, (d) whitewashed, or (e) left as the bricklayer leaves it.

639

This will prevent the bricklayer from doing unnecessary work and will also enable him to lay stress on that class of workmanship that is best suited for the purposes of the particular finish used.

Fig. 133.—Two Man Straight Edge for Jointing.

640

For example:

(a) If the plaster is applied directly, the joints must be left ragged and raked out deep.

(b) If the wall is to be furred, no jointing nor striking is necessary, smoky brick can be laid on the face. Provision for nailing must be made.

(c) If the wall is to be painted, work should be jointed or struck, not only for looks, but also to collect the minimum amount of dust. All holes and pits in either brick or joints must be thoroughly filled with mortar.

(d)   If the wall is to be whitewashed, the whitewash can be put on thick enough to fill all pits, Jointing and striking are not so necessary.

641

All jointing must be done in the following manner: The joint must first be smoothed so as to remove any distinct line where the edge of the brick leaves off, and where the mortar of the joint begins.

642

After the joint has been made smooth and flush with the face of the brick, the joint must be rubbed with a jointer close to the top of the edge of the brick.

Fig. 134.—One Man Straight Edge for Jointing.

643

The jointer must be used with a straightedge about 8 ft. long, ⅞ in. thick, and 3 ins. wide, preferably used by two men at the same time.   (See Figs. 133 and 134.)

644

The jointer must make only a slight impression in the mortar, and must not be pushed in to a point where it discloses the variation in thickness of the joint, unless called for in the specifications.

645

A straightedge must always be used for the reason that it not only makes better appearing work, but the work is much cheaper to do.   The straightedge, of course, is not used for the vertical joints.   These should be ruled even less deep than the horizontal joints.   Always complete the vertical jointing before ruling the horizontal joint.

646

The finish of the joints on the exterior face varies under different conditions.

647

Figure 135 shows the seven most common methods of finishing joints.

(A) A joint made in this manner will appear smaller than it really is. Where the mortar is not noticeably different in color from the color of the brick, this is probably the best way of hiding the inaccuracies of thickness of the different brick in the same course.

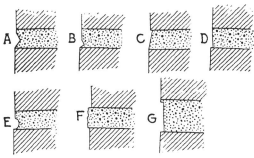

Fig. 135.—Method of Finishing Joints.

(B) This finish is similar to finish (A), except that it is deeper and makes the joint look much wider than does finish (A).

(C) This style of finish is made by holding the handle of the trowel down below the joint, and smoothing the joint down with the blade of the trowel. It is supposed to shed the rain much better than finish (D), but in practice it is probably but little better than any of the other finishes, especially when plenty of good Portland cement is used in the mortar. The main objection to this method of finishing is that it takes the eye away from the top edge of the courses of brick, which are the real straight lines, hides the accuracy of these lines, and exposes and exaggerates the inaccurate lower edges of the bricks of different thickness, by accenting their shadows on the sloping surface of the joint.

(D) This style of finish is made by smoothing up the surface of the joint while holding the handle of the trowel up above the joint. Theoretically, this finish does not shed the water as well as does finish (C), but it requires less skill to make, and it hides the uneven lower edges and accents the straight and even top edges of the brick, as it leaves no shadows at all.

(E)   This finish is made by a half round jointer. This method of finishing can be done quicker than either (A) or (B), but it does not look so well for most work.

(F)   This style of finishing is used with coarse sand where certain textures of wall surfaces are required. The effect is secured by giving the brick a tap with the trowel after the mortar has been cut off.

(G)   This method of finishing is used where wide joint effects are desired. The work is done with a jointer. For the best effects, it should be done exactly as shown, i. e., with the jointer resting on the top edge of the brick and with the projecting mortar hiding the lower edge of the brick.

648

When brickwork is painted and striped to represent the joints, the best results can be obtained by locating the stripe on the face of the brick adjoining the top edge to represent the horizontal joints, and on the face of the brick adjoining the end of the brick to represent the end joints.

649

If the striping is done as directed, the inaccuracies of the wall will apparently be decreased. If the stripe is located on the mortar joints, the roughness of the workmanship of the wall will apparently be increased.

650

The following process must be followed when pointing brick work:

(a)   Drench thoroughly with a hose for several minutes until the brick and mortar will not soak up any more water.

(b)   Apply strong muriatic acid until the particles of sand and the surfaces of the brick are cleaned of cement.

(c)   Wash thoroughly with a hose under high pressure until all traces of the acid have been washed off.

(d)   Rub in a very thin layer of neat Portland cement into the surfaces of the sand and brick.

(e)   Immediately put in the new pointing. This process will make new pointing adhere to old as firmly as if all were put in at one time, as is proved by many pieces of work that have stood the test for fifteen years.

651

The entire secret lies in getting absolutely clean surfaces of sand or brick or stone to adhere to and these clean portions act as dowels between the old work and the new.

652

Care must be taken, when using acid, to keep ropes away from it or its fumes, as a few drops of the acid will ruin their strength.

653

Keep falls out from under the staging, or they may be injured by the acid and the workmen may get hurt.

CHAPTER XVII.

ARCHES AND CHIMNEY BREASTS.

654

The cutting of brick arches in the shanty on wet days should be used as a reward to the most faithful bricklayers, and those who have operated most nearly in accordance with this system even if it is not more economical than cutting the arches on the scaffold.

655

Make the detail so as to require the least amount of cutting possible on each brick.

656

Where there are a large number of arches to be cut to the same pattern, it will be found economical to make the pattern out of sheet metal instead of paper.

657

All brick work to be cut should be laid out on stiff thick paper by the foreman. After it is laid out, with what represents each brick numbered and tacked to a board, the joints should be cut out with a sharp knife.

658

These boards, with each paper brick nailed in place, should then be given to the bricklayers, who should draw the tacks, mark out the shape of each paper on a brick, number the brick the same as the paper pattern, and then tack the pattern back on the board.

659

The brick, as fast as they are cut, should be laid on the floor face up.

660

When the arch is completed, it should be packed carefully in hay or straw into a box or barrel.

661

The barrel should be marked with the name of the bricklayer, the number of the arch, and the time it took to cut it.

662

Do not have anyone but a foreman or a first class bricklayer lay out the arch on paper, or the arch will cost more done this way than cut on the scaffold.

663

Do not pack more than one arch in one barrel, or the cost of sorting the brick will be excessive.

664

Do not take the brick out of the barrel until the barrel has been placed near the place where the arch is to be laid.

Have the bricklayer who cut the arch lay it.

665

The foreman mason must write on the blue prints of the elevations the names of the bricklayers on the various portions of the work that they build. This will enable the foreman to locate the blame for a soffit that looks badly when the center is taken down.

666

The moral effect is also in the interest of good work.

667

Extra speed also can be obtained by letting the bricklayer know that his name and the hours and minutes it took him to build this arch are written down on the blue print.

668

The greatest amount of labor on cutting a straight arch is expended on the soffit or bottom face.

669

An arch straight on top and curved underneath, requires no cutting of the surface underneath. It must simply be split off to a line, to the right thickness.

670

This requires very little labor, and the natural surface of the brick forms the soffit.

671

The more flare to the skewbacks, that is the more they slant from the vertical, the more courses of brick in the arch, and the more difficult the cutting of each individual brick.

672

The more the skewback flares, up to reasonable limits, the better the arch looks.

673

The less it flares, the less it costs to cut the brick and to lay them.

674

There is no arch so cheap to cut and to lay as the one

where the joints are radial to the curve of the soffit. (See Fig. 136.)

675

The more camber to a straight arch, the less cutting to the surfaces of the brick that rest on the centering.

676

Give the top line of a straight arch a little camber, but not nearly as much as the soffit, for it is apt to settle a little when completed, and it always appears to the eye to be a little sagged anyway.

Fig. 136.—Building Brick Arches.

677

The rubbing of the soffit of brick arches should not be necessary, if the bricks are cut properly.

678

If they are to be rubbed, the brick should be put against a carborundum wheel, as it will do the work quicker than any other method.

679

A rubbed brick will never, at best, look any better than a piece of sandpapered wood carving.

680

Foremen must see that the angle of the skewback is made identical to that shown on the architect's plans.

681

No alteration shall be made unless ordered by the architect in writing.

682

It is of importance that all skewbacks of similar arches shall be made at exactly the same angle from the horizontal.

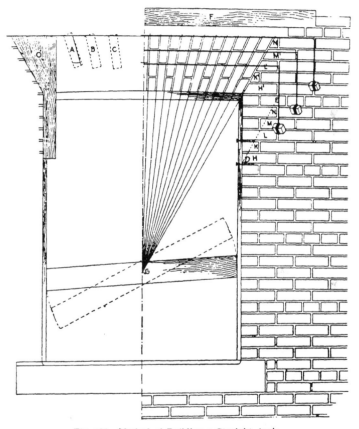

Fig. 137.—Method of Building a Straight Arch.

683

As bricklayers' time costs too much to have them making measurements, the foreman must provide them with wooden skewback patterns, like O in Fig. 137, with all the courses marked on them.

C84

Supports for wooden centers for straight arches can be best held in place by boards or upright joists resting on the sill, as shown in Fig. 137.

685

The uprights can be held in place by a spreader, G, and by nails driven into the brickwork. The nails will hold better in the green mortar of the jamb if they are not driven clear up to the head. Fig. 137 also shows places where the lines for ranging the joints should be strung.

Fig. 138.—Rowlock Arch Bonded by Rings.
(Note that crosses indicate real headers.)

686

The best results can be obtained by giving the center a slight camber of say 1/16 to ⅛-in. rise for each foot of span of the opening, or the arch will appear to the eye to be sagged.

687

All joints in the first line of joints above the soffit should be the same distance above the wooden center, or the camber will be noticeable.

688

Each course of brick should be the same width on the top, measured on a line perpendicular to its radial joints, and not on the level line of the top of the arch.

689

Each course should be cut to the full width of the brick placed in the positions of A, B and C.

690

It is a common practice to space off evenly the courses at the top of the arch, but this costs more and does not look

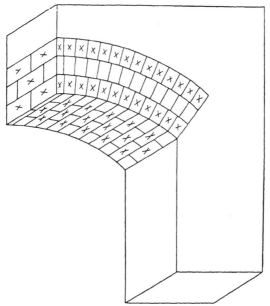

Fig. 139.—Rowlock Arch with Bonded Soffit, and Bonded by Rings.

as well as when each course is spaced by marking it off with the full width of the brick (plus joint) held in the positions as shown by A, B, C, and then marking off the bottom joints on the wooden center by radial lines extending from the upper marks to a center found by extending the lines of the skew-backs. These lines reprsenting the joints should be extended up to a straightedge, F, placed one course higher than the top of the arch, so as to be out of the way of the top brick that are being laid.

691

The bricks to be cut and laid adjoining a skewback can be measured and marked most quickly if a line corresponding to and parallel with the skewback is marked upon the finished wall as shown by the dotted line DE on Fig. 137.

692

The line must start from an even numbered or corresponding course below and of the same bond as the first course of the skewback.

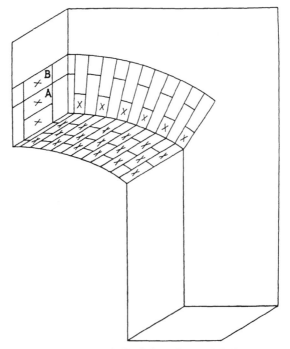

Fig. 140.—Arch Bonded on Face and Soffit.

693

This line will enable the bricklayer to take the measurements instantly along the top edge and bottom edge of the corresponding brick to be cut. H' can be marked for cutting to correspond to H', K' to K, etc.

694

All arches must be thoroughly bonded.

695

If the soffit is not to show at completion, it shall be built similar to Fig. 138, i. e., with all stretches in the lower ring. The next ring should be all bats, all stretchers and all bats; the third ring should be all stretches, and so on.

696

If the soffit of the arch is to show at completion, as in Fig. 139, the lower ring must be bonded as shown, and the 2d and 3d rings must be bonded in the same manner as Fig. 138.

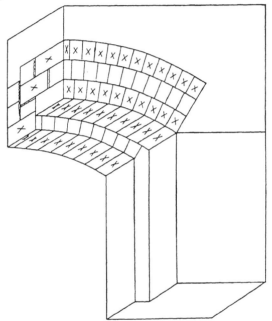

Fig. 141.—Rowlock Arch Over Window Frame.

697

This makes a better looking arch than Fig. 138, but it is no stronger.

698

If the face and soffit are both bonded, the arch should be built similar to Fig. 140, and the filling at A and B should be all rowlock headers.

699

These rings do not need to be bonded into the face of the arch, as the regular headers of the wall will tie it all together—furthermore, whole brick unbonded will be stronger, and will also be much cheaper than brick cut to fit the brick on the face of the arch.

700

If the lower ring of a rowlock arch is laid on top of wooden lintels that will remain in place, the lower ring does not need to be bonded in itself, but instead can be bonded by the ring over it. See Fig. 138.

701

Arches over windows seldom come on the same level clear through the wall, in which case the arches, whether bonded on the face or not, should be built similar to Fig. 141.

702

The rings are bonded on each other instead of in themselves.

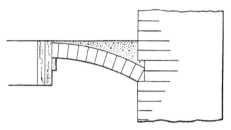

Fig. 142.—Details of Trimmer Arch.

703

The lower ring on the outside should be leveled up to receive the inside rings, which should be laid out to and against the back of the outside arch.

704

On account of the shrinkage of wooden joists, it is advisable to start the trimmer arch as low as possible, and to end it as high as possible. In cases of this sort, the arch must have a slot for forming the skewback, instead of a corbelling, as the latter will interfere with the furring of the story below.

705

A set-in 1 in. deep 2 courses high is all that is necessary to hold any 4-in. trimmer arch. (See Fig. 142.)

706

When making chimney breasts, do not forget to set in two courses of brick the entire length of the chimney breast to form the skewback of the trimmer arch. See Fig. 142.

707

When flues are carried over to one side it is necessary to cut and fit the flue lining. This requires considerable skill.

708

Small sizes, made of soft material, can be cut with a hammer and chisel if the pipe is placed vertically on end and filled with sand.

709

The large sizes of the hardest of vitrified lining can best be cut by standing on end and striking downward with a bricklayer's hammer. Strike until the pipe is made thin, then break off the tops of the thin sections with a hammer. The downward blow tends to crack the pipe least. Pipes can be cut quickly in this way without cracking.

710

Fireplaces should be located as nearly in the center of the chimney breast as possible. The top of the fireplace should be brought to the center evenly from each side, until the size of the flue is reached. The flue should then be built toward that side of the fireplace above that has the fewest number of flues on it. This will assist in locating the fireplace above in the middle of the breast.

711

The flues should have no flat places that will collect mortar dropped from above while building, nor collect soot after the fireplace is used.

712

At every bend in the flue, leave an opening at least 2 ft. high for a board that will permit the mortar dropped from above to fall out on the floor. If the flue has clay flue lining fewer clean-out holes will be required.

713

Flues not lined must be plastered smooth.

714

All joints in the brick work of chimneys must be filled absolutely full of mortar.

715

All withes must be tied into breasts at least every five courses.

716

Be extremely careful to fill all joints between headers on chimneys, as expansion and contraction will sometimes permit smoke and fire to escape in the straight joint between headers.

717

All brickwork on chimneys and chimney breasts must be kept away from the wooden floors. Extra precautions must be taken to fill all the joints and to plaster up the inside of all withes, so that no fire or heat can come through and start a fire in the dry wood surrounding it. There should always be a space at least 2 ins. wide between the brick and the wood.

CHAPTER XVIII.

TEARING DOWN, CUTTING OUT AND PATCHING BRICKWORK.

718

Wherever possible, walls to be taken down should be thrown over in large sections, instead of tearing down a small section at a time, because the bricks will be in better condition. (See Fig. 143.)

Fig. 143.—Wall Thrown Over in Large Sections to Reduce Breakage of the Brick.

719

When large quantities of brick are to be taken down, as on the eight story building that we took down for the Mutual Life Insurance Company at San Francisco, wooden chutes are the most economical. (See Figs. 144 to 146.)

720

Arrange these chutes so that their direction will change at least once per story, or else put in wooden baffle plates hung loose from the top side of the chute, so that the bricks will not travel fast enough to break up.

721

At each floor level provide an opening on the top side of the chute for relining the bottom as fast as it wears thin.

181

722

The more care that is taken in designing the chute, the more bricks will be in condition to be used again.

Fig. 144.—Wooden Chute for Conveying Brick from Demolished Walls.

Fig. 145.—Wooden Chute for Conveying Brick from Demolished Walls.

723

When cutting out old brickwork, the greatest speed can be obtained by using a short chisel made with an octagonal shaped shank.

724

The brick can be cut out the fastest if the chisel is put on the lines shown in Fig. 147, and if the pieces are removed in exactly the order numbered in this figure.

725

If a large hole is to be cut out, the greatest speed can be obtained by cutting out one complete course, with each brick removed in the manner shown, and then proceeding on the next course above.

Fig. 146.—Wooden Chute for Conveying Brick from Demolished Walls.

726

It is, obviously possible to work fastest by cutting out from the bottom, as the pieces will then drop as fast as they are loosened.

727

If the cutting is done from the top downwards, the pieces must be removed as fast as loosened.

728

When second hand brick are used, the greatest care must be used not to lay brick that have any smoke stains on them on the line.

729

If such brick are ever whitewashed, the smoke will stain through and give a bad appearance to the wall.

730

The purpose for which a wall is to be used should determine whether or not all of the interior vertical joints in the wall should be entirely filled with mortar. Where dryness,

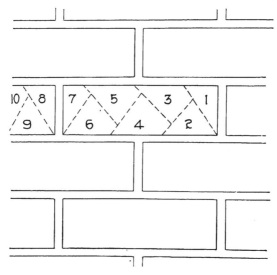

Fig. 147.—Method of Cutting Out Brickwork.

least condensation on the inside, and general serviceability are required, the best method of bricklaying for exterior walls of buildings is what is called the "brick and brick" method. In this method the bricks in the middle of the wall are laid with the stretchers touching each other end to end, and the headers touching each other side to side, with no attempt to get shove joints. There is no special attempt to fill the vertical joints by shoving except on the outside face tier and the inside face tier.

731

The brick on the interior of the wall must be laid in soft mortar, the brick being so wet and the mortar so soft that the mortar will run well down into all the spaces.

732

We realize that this is revolutionary, and against all old-fashioned doctrines; but it must be remembered that we are building under conditions today that are distinctly different and that upset many of the old traditions in brickwork.

733

We are using better mortar today than ever before. The low price of our best cement removes all excuse for using any mortar that will not be much harder than the brick with which it is used.

734

The modern use of hollow brick has taught us that a certain number of air spaces in a wall are advantageous in keeping out moisture. We now look with pity on the man who used to create a disturbance if he found a space between two bricks that lacked one thimbleful of mortar, while he would not hesitate afterward to cut through the wall and make a window opening five feet square in the same place.

735

The exterior tiers on the exterior wall must be completely filled with mortar, but on the interior tiers it is practically impossible to get shove joints in each and every case.

736

Specifications on our best work have called for shove joints for years, though it is a well-known fact among bricklayers that it is next to impossible to shove each and every joint full of mortar.

737

Furthermore, when the fingers of the bricklayers get worn so thin that they are nearly ready to bleed, as is often the case, it is still more difficult to get them to shove the brick into the mortar. Then comes the practice of slushing up the unfilled joints with trowels full of mortar. While joints so filled appear full of mortar, the work is not properly done. All this finally results in the brick being laid the width of a joint of mortar apart for slushing.

738

Laying the interior stretchers so that they touch each

other end to end, and the interior headers touching each other side to side, results, in actual practice, in getting about ten brick in the wall where nine brick with shove joints would be.

739

The wall will, under actual working conditions, have a much smaller amount of unfilled voids. The brick will have a greater length of lap or bond, on the brick above and below.

740

In freezing weather especially, when shove joints should never be used on account of the certain heaving of the freezing mortar, this is undoubtedly the ideal way to lay brick.

741

Furthermore, the bricks will be laid exactly as planned, and it will be easier to obtain the desired arrangement of cross joints.

742

It is the easiest way for the bricklayer, and he is doing honest work. With shove joints he is not always able to do what he is expected to do, and he lays the brick a wide joint apart. Then, under certain conditions, these joints may be "slushed," which will prevent the next brick from pushing the mortar into the unfilled joints of the course below, and there will be about 10 per cent. less brick in the filling at the finish.

743

Walls built as above described may be plastered sooner, with less danger of uneven discoloring or spotting. A wall so built will dry out much sooner. There are important features for the comfort of the tenant or the operative who is to move in the day the building is finished, as is the case with our American office buildings, hotels, apartment houses and factories.

744

We realize that many intelligent people will not understand and believe that the "brick and brick,' or "brick-touching-on-their-vertical-joints," method is the best way to lay brick on buildings, but every bricklayer knows this method and approves of it, provided good soft mortar or grouting is used.

745

Knowing the general prejudice in favor of shove joints, we would not advocate the "brick and brick" method, if it were not for the fact that we cannot be suspected of any other motives than that of saving money for the owners for whom we work, as all of our work is and will be done on the basis of cost-plus-a-fixed sum only, and under this form of contract all savings go to the owner.

746

And what are the disadvantages of the "brick and brick" method? None, except the prejudice against it held by those who have been taught, without any real proof or reason, that shove joints (which we do not really get except on a part of each bricklayer's work), are the best.

747

The foregoing applies only to building construction. Do not misunderstand it. This method does not apply where absolutely filled joints are required, such as on hydraulic work, sewers, and on work subject to alternate wetting, freezing and thawing.

748

On sewer and chimney construction, and on all work that does not require extreme accuracy, and where there is no danger of shoving the face work out of line, no other method than shove joints should be tolerated.

749

When large horizontal joints on face work are required there are three ways of obtaining them:

(a)   By using stiff mortar.

(b)   By using mortar of very coarse sand.

(c)   By bedding house slate or thin tile back in the joints.

It is not possible to make hard and fast rules on this subject, as conditions demand sometimes one method, sometimes another.

750

Joints made by the third process will certainly shrink the least. This is often an important consideration.

751

Until very recently, it has been the custom to consider that the smaller the joints the better the work.

752

Modern mortar composed of Portland cement has changed all such ideas. It is rare to find a brick to-day that is as strong as good mortar made with a fairly large proportion of Portland cement, and Portland cement is so cheap to-day that there is no excuse for not using it.

753

Therefore, the size of the joints should be increased on common work until there is no question as to the quality of workmanship of the bedding of each and every brick.

754

Every inspector knows that with tight joints the great difficulty is to get brick bedded in such a manner that, if the laid brick should be lifted, it would not still show a hollow or a mark of the trowel in the mortar.

755

There is therefore no danger from a strength standpoint in getting too large joints, if the mortar is stronger than the brick.

756

On certain kinds of face work, the desired effects of texture cannot be obtained except with unusually large joints, especially bed joints. The architect's orders in this respect must be executed literally.

757

The foreman must see that the mortar furnished for this work is stiff enough and coarse enough to give the desired thickness of joints without running or sagging from the bottom edge of the brick above the joint.

758

If local conditions are such as to make it difficult to obtain joints of the thickness required by the architect, small pieces of tile or house slate must be bedded in the joints back far enough not to show, but sufficient to give the architect exactly what he wants without any substitutes or excuses whatever.

759

The greatest care must be taken in measuring the ingredients of each batch of face mortar especially for wide joints, or the wall will have a spotted effect due to different color which is caused by varying the proportions of ingredients.

760

It is also very important that no part of wide joint work dries out quickly, or it will be permanently of a lighter color than that mortar which dries out slowly.

761

It is generally good practice to lay up several 4 in. samples of face brick work about 3 ft. long and 5 courses high, jointed in several different ways, for the architect's inspection, selection and approval before laying brick on the wall.

762

After the sample is approved by the architect, see that all bricklayers on the face work are provided with the same kind of jointers.

763

Keep the sample where they can all see it, in order to have all of the work as uniform as possible.

764

All face brick work should be backed up at least 4 ins. before it is left at quitting time, or the mortar will surely dry out a lighter color, and its strength will be much decreased.

765

If it is not possible to back the face brick work up 4 ins. before quitting, plastering the back of the face brick ½ in. is the next best protection.

766

If brick is laid with terra cotta, the laying of the terra cotta, if it is combined with brick, is generally considered to be bricklayer's work.

767

All portions of terra cotta blocks inside the face of the wall should be filled solid with bricks and mortar or concrete at least as strong in compression as the terra cotta itself.

768

The filling of the holes in the back of terra cotta should be done only when the terra cotta is standing on its vertical face.

769

If the terra cotta varies sufficiently in thickness in a direction crosswise of the walls, or say at least 4 ins. from the courses under and above it, sufficient lap will be furnished by the terra cotta itself.

770

But if it does not afford sufficient lap for the headers above or below it, then the holes in the back must be filled with brick.

771

The brick should project at least 4 ins.

772

Do not fill the terra cotta until it is about ready to set on the wall, or the terra cotta will probably be broken in handling.

773

Do not attempt to lay the terra cotta first and to fill the holes as it is backed up, unless you deduct the entire width of the terra cotta from the working thickness of the wall.

774

Do not fill that part of the terra cotta that projects beyond the face of the wall, unless to keep out snow and rain.

775

Terra cotta shrinks so much, and so unevenly in the baking that it seldom lays evenly to the marks made for it.

Fig. 148.   Fig. 149.
Correct Way of Shaping Brick for Filling Putlog Holes.

776

It is generally better practice first to lay an entire course on the wall dry and to mark the joints on the course underneath, then to roll it over on its back, bed it, and roll it forward into the mortar.

776a

A putlog hole must be filled in the following manner:

(a) The interior of the hole must be dampened.

(b) The hole must be filled with more mortar than is actually required to fill the joints, and the mortar must be plastered on to each face of the hole.

(c) The brick must be cut wedge shaped on the back. (See Figs. 148 and 149.)

(d) The brick must be hammered so that the mortar is squeezed out on all joints. (See Fig. 150.)

(e) Use the same mixture of mortar as on the rest of the wall.

(f) If the hole is not solid full of mortar and filled in the manner described, the mortar will dry out lighter in color than the rest of the surrounding wall.

(g) Do not do the jointing better than the adjoining jointing. If the surrounding jointing is bad, do not try to correct it all on one brick. Make the jointing around the one brick just as bad as that which adjoins it. This applies also to all patching and toothing.

Fig. 150.—Correct Method of Filling Putlog Holes.

777

Clear out all putlog holes and toothers before mortar sets hard. It is much cheaper to clear it out while it is soft. In the case of toothers, cut out the temporary bat before the mortar hardens, as the toother is apt to break off during the cutting out of the bat, if the mortar is hard.

778

When laying the end of a new brick wall with a straight joint up against an old brick wall, use no mortar in the joint between the new and the old face brick, or the shrinkage will surely show a crack in the straight vertical joint.

779

When it is desired to increase the thickness of an old wall by lining it up with a new wall, many different methods may be employed for bonding the new work to the old.

780

Undoubtedly the best way, and the one that will give a perfect result without any iron ties, is to cut pockets in the old wall, as shown in Fig. 151, from the top of one header course up to the bottom of the next header course above.

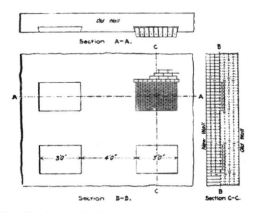

Fig. 151.—Pocket Method for Joining New Walls to Old.

781

These pockets should be one tier deep, 3 ft. long and 4 ft. apart on blank walls. They should be splayed in on their vertical edges so that the bricks will dovetail in. All old mortar must be cut out clean, unless it is considerably harder than the brick to which it is attached. In any case, all shelving surfaces should be cut down level.

782

As soon as the new lining wall is built up as high as the bottom of the hole, the entire hole is to be bricked up as shown, with nothing but the hardest of headers, laid header over header with the softest of mortar, and with the thinnest of joints.

783

Each brick should be pounded inward and downward, the top course being thoroughly wedged off with brick and house slate.

784

Do not have any wedging project that will interfere with the header course of the new wall getting a good bond on the top course of the new block.

785

Many bricklayers build these blocks with every other course a header. Bonding out to the face of the new wall every other course with headers of these blocks is wrong.

786

The purpose of these blocks is to form a ledge on which to put the regular headers of the lining wall. This ledge is sure to carry the most of the weight of a large part of the lining wall, unless the lining wall is laid up slowly and with very close joints.

787

Unless the block are all headers and thoroughly wedged and dovetailed, they will either shear off or draw out.

788

Lining wall should generally be laid with close joints to reduce the amount of shrinking.

789

A cheaper method, but one not so good, consists of using bolts extending through the old wall and the new lining. Large washers under the head and nut of the bolts keep the lining from splitting from the old wall.

790

On general principles never do any toothing unless it is absolutely necessary, because of the difficulty in getting first class work on the filling of the toothers.

791

Blocking is much stronger than toothing.

792

Consequently, if it is necessary to tooth, tooth the outside tier only.

793

The interior 4 ins. or tier must be blocked instead of toothed. (See Fig. 152.)

794

When building out for blocking, always corbel with a ½ stretcher on all interior tiers at once, and always set back a ½ stretcher at a time. (See Figs. 152, 153 and 154.)

795

If any other bond than common bond is used on a wall one portion of which is built up before it is all built up, toothing should be used, but if common bond is used and the **greatest** strength is required, blocking can be used to great **advantage.**

Fig. 152.—Toothing Face Tiers, Blocking Filling Tiers. Fig. 153.—Racking Split Blocking.

796

Do not carry the bond around the end of the blocking, as this weakens the block.

797

The more stretchers used the better the new block is tied to the old block. (See Figs. 154 and 155.)

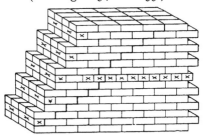

Fig. 154.—Racking Straight Blocking.
(Note the distance saved on the end of the wall by blocking racking instead of course racking.)

798

Always run the stretchers out on every other course to show their ends on the end of the block. The courses in between must have brick with their sides on the end of the block. (See Fig. 154.)

799

Bear in mind that the off-sets on the blocking of the wall that is built first will have to carry more than their normal

800

load. The first wall will have completed its shrinking before the second wall is built, and as the second wall is built it will shrink and throw extra weight on the blocking of the first.

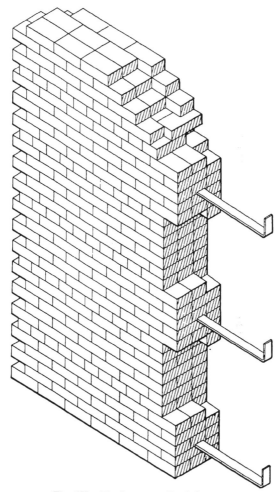

Fig. 155.—Tie Irons, or Band Irons.

801

The top of all blocking must be cleaned off so that the new work will get a good hold, and a good bearing.

802

The new blocking must be made of stretchers that run into and against the old blocking every other course.

803

If blocking or toothing must be used, a liberal supply of tie irons should be built in at least 2 ft. each way to hold the walls together while shrinking and setting. (See Fig. 155.)

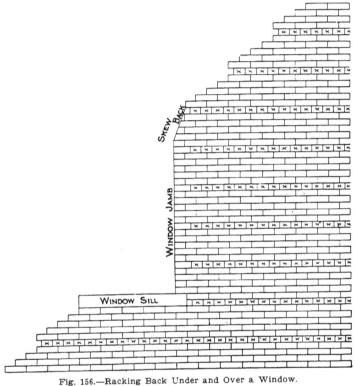

Fig. 156.—Racking Back Under and Over a Window.

804

If blocking and toothing are practiced only where absolutely needed, it will soon be noticed how seldom they are really necessary.

805

There are many ways to do without blocking, even when it is desirable to build one wall up a story or two higher than another, as follows:

(a) Racking back. (See Fig. 156.)

(b) Racking back and toothing only over the header course. As the toother over the header is only one-quarter of

Fig. 157.   Fig. 158.
Correct Way of Shaping Brick for Filling Toother.

Fig. 159.—Correct Way of Filling Toother.

a brick long, it can be easily filled with mortar under the pressure of driving the header in place. (See Fig. 156.)

(c) Racking back under and over a window, taking advantage of the straight line of the window jamb and the skewback. (See Fig. 156.)

806

If toothing or blocking must be used, do not have the toothing or blocking any higher in a plumb line than is necessary, that is, either rack back a part of the way (see Fig. 155) or set over on a new line several feet away.

Fig. 160.—¾ Toothers.

Fig. 161.—San Francisco Band Irons.

807

In filling a toother, cut the back portion of the top and bottom sides of the brick (see Figs. 157 and 158) so that the brick forms a wedge when it is driven diagonally toward the back corner into the pocket of the toother.

808

This pocket must be previously completely filled with mortar which has been rubbed hard into all its surfaces, top as well as bottom.

809

The brick must be plastered on top, bottom and end before driving into the pocket. (See Fig. 159.)

810

Figure 160 is an example of "¾ toothing." In some cases this is the best method of toothing, because the permanent bat actually carries its proportional weight, and because the tooth pocket on the new wall, being only ¼ of a brick deep, can easily be completely filled with mortar.

811

Figure 161 illustrates the conventional method in San Francisco of tying walls all the way round a building at all floor levels by using 4 x ½-in. bars with interlocking hooks at the end. Extra sets of bars are employed where toothing occurs, as shown.

CHAPTER XIX.

BOND.

812

The word "bond" has two meanings:

(a) From a pattern standpoint, bond means the relative arrangement of the vertical joints.

(b) From the standpoint of strength, bond means the amount of lap one brick has on another, to tie two bricks together.

813

No brick should ever be lapped less than one-quarter of the brick on which it rests.

814

No closer should ever be less in length than a quarter of a brick.

815

Bricks, terra cotta, stone and other blocks are imbedded in mortar for four reasons:

(a) To bring their tops to the right height, or level.

(b) To provide a method of distributing the weight evenly.

(c) To cement the bricks together.

(d) To keep out air, rain, water or moisture.

816

Every effort should be put forth to make the bricks and mortar adhere as of one mass; but the real purpose of bond is to make the bricks stand up and carry their weight even though there is no adhesion between the brick and the mortar.

817

All walls must be "laid out to bond" before starting the actual laying of the brick in mortar regardless of the kind of bricks that are used and of the place where they are used.

818

If the bricks are common bricks of "run of kiln" quality, they will vary in length considerably.

819

A course of brick should be laid dry on the wall in order to make sure that the different bricklayers start their pieces to the common bond of the wall. Otherwise there may be a small piece of brick required for the closer between the work of each two men on the wall.

820

It takes about three times as long to cut and lay a closer as to lay a whole brick for a closer.

821

Therefore, even on rough inside work, the bond must be laid out dry for the sake of saving the time necessary to cut the closer on each course.

822

One whole course must be laid on the wall.

823

It is not enough to lay three or four brick, then pass them along, marking them as they are laid over and over again. Bricks vary so much in length that the only safe thing to do is to lay out one whole course.

824

The bond laid out on a straight wall is not affected nor changed by any pilaster, regardless of its amount of projection or shape of plan, provided, (a) that all angles and corners of the pilaster are square; (b) that the pilaster has no brick less than full size in it; and (c) that the bricks are of correct proportional measurements. See Bond Chart No. 2.

825

It is, however, safer to lay out the bond as running in the straight line of the plane of the main wall, and also to lay it out to fit the entire perimeter of the pilaster.

826

Any openings or breaks in the wall must be laid out to show the way the bond will work out over the top of these openings.

827

If some one of the accepted patterns will not work out then several stretchers must be clipped, to shift the bond gradually.

828

Do not, under any circumstances, shift the openings or breaks in the wall to fit the bond, unless ordered to do so by the architect in writing. Even then, due consideration must be given to the effect on the trim around the inside of the openings.

829

The laying out of the bond on plumb bond work (i. e., absolutely plumb joints) requires a great deal of judgment, and must be entrusted only to the bricklayer who is known to be able to do it properly.

830

The foreman must, in any case, check up the layout before the men have laid more than two courses.

831

In laying out the bond on any wall where the bricks are culled for color, due consideration must be given to the fact that the longest brick will be used at the bottom of the wall and the shortest brick will be laid at the top of the wall.

832

This is necessary because the brick that are burnt the least will not only be the largest but also the most perfect in shape, because the longer they are burnt the more they will shrink in all dimensions.

833

Those brick that are burnt the longest, being at the top of the wall, are so high above the ground that their inaccuracies in shape will never be noticed.

834

Therefore, it is not safe to proceed by the layout of the first cull of brick on the wall until a course of the last cull is also "laid out" on top of the brick of the first cull, to see how much difference there will be in the length of the brick and the size of the end joints.

835

With some makes of brick, this difference is so great that plumb bond cannot be carried. Then the bond must either be started new over belt courses, or projections, or have courses of full headers introduced at least as often as the culls change, or else not use all of the culls on any one wall.

836

Oftentimes different culls can be used on different walls, but there is then generally difficulty in getting the two walls to lay up alike, as the bed joints on the smaller culls must be made much larger to be level with those built of the larger culls.

837

In any case, the following rules must be observed when laying out bond:

(a)   Lay out dry the entire length of the wall, at least one course of the culls that will be used at the bottom of the wall.

(b)   Lay out dry the entire length of the wall at least one course of each of the upper culls.

(c)   Lay out jambs of all openings above the first course as if they occurred in the first course. Shove the bond back until it fits the jambs of the openings, unless the 2-in. piece will make the bond fit the jamb.

(d)   Shoving the bond back will require one or more ⅞ brick for stretchers in the walls above and below the openings. These will do more good than harm. They will never be noticed, and may become very welcome if, for any reason, the bricks actually used happen to be shorter than those with which the bond was laid out.

(e)   In other words lay out the bond of the upper part of the wall before you start laying any of the wall. Then when you carry plumb bond you will not have clusters of short brick in the wall from top to bottom or any undesirable bond around jambs, corners, angles, or at the junctions with other work.

838

On short lengths of wall that do not lay out to full bond (i. e., requiring a three-quarter brick), start from the angle with a three-quarter. Or, in other words, run the piece into the angle.

839

There is no way of bonding brick to a backing that is so effective as with real brick headers that show on the face as headers.

840

Figures 162 and 163 show the interior court of the St. Francis Hotel, at San Francisco. The entire face of this court, 8 stories high, peeled off by buckling, due to expansion by the heat of the great fire. These figures show how valueless the "blind" (clipped diagonal) header is for bonding face brick to the backing.

841

Galvanized iron ties should always be used in connection with blind or clipped brick headers.

842

The metal ties will hold in case of buckling due to heat, and the clipped headers (see Bond Chart No. 2) will help to hold in case the metal ties rust off as a result of not covering the ties completely with mortar.

843

There are many kinds and patterns of anchor irons. Those that are the least expensive to brick in are those that bend up only. If anchor irons have projections that bend down they are harder for the bricklayer to build in.

Fig. 162.—Court of St. Francis Hotel, San Francisco. After the Great Fire.

844

In building in iron anchors, be sure to lay the brick as hard against the anchor as possible, so that it will have no loose motion or play whatever.

845

The building laws of many cities call for bond stones in brick piers.

846

These must be avoided on our work wherever possible, as recent fires have shown them to be very dangerous, due to the disintegration of the stone by heat.

847

The various isometric drawings and transverse sections in this system are drawn with intention of showing the appren-

Fig. 163.—St. Francis Hotel, San Francisco, After the Great Fire.
(Note the diagonal headers.)

tice the bond of the various tiers in themselves and to the adjoining tiers. The general features of bond will be better understood, and the individual drawings will serve their purpose better if the following rules are carefully studied:

(a)   A wall consists of three different kinds of tiers: (1) outside face tier, (2) inside face tier and (3) filling tiers.

(b)   All interior or filling tiers should have the end joints of the stretchers on a straight line with each other and in line with the end joints between the stretchers on the outside face tier as in Bond Chart No. 1, if the outside face tier is of the same kind of brick and will bond with the filling tier.

(c)   If the filling tiers, for any reason, cannot be made to bond with the outside face tier, they should be made to lay in a straight line in themselves across the wall and in line with the end joints between the stretchers of the inside face tier. See Bond Charts No. 3, No. 4, No. 5.

(d)   If the filling tiers, for any reason, cannot be made to bond with either the outside face tier or the inside face tier, then the filling tiers shall be laid with the end joints between the stretchers in a straight line with each other, and special care shall be made to see that the joints break evenly over the center of the stretchers of the course immediately below the course being laid. This will require extra precautions, as the course below is covered with mortar to receive the next course. While, if the stretchers will lay to bond with either the outside face or the inside face tier, the breaking of the joints of the course below on the filling tiers will be automatic, and will therefore, require no watching of the location of the joints of the course below.

(e)   If the wall is an even number of tiers thick, and the outside face, filling, and inside face tiers all bond with each other, that is, are all laid with the same size of brick and same size of joints, then the end joints between the stretchers should run through the entire thickness of the wall in a straight line.

(f)   If the wall is an uneven number of tiers thick, and the outside face, filling and inside face tiers all bond with each other, then the end joints of all stretchers of the filling and outside face tiers should be in a straight line and the end joints of the stretchers in the inside face tier must be located opposite the center of the stretchers of the filling tiers, in order that the bond will lay out properly on the jambs of any openings that may occur, or be afterwards built in the wall.

If the wall is laid differently than described in rule "e" or rule "f," the bricks on the jambs will require a lot of unnecessary cutting.

(g) If the wall is an even number of tiers thick, whether or not the bricks on the various tiers are of the same thickness, have the headers throughout the entire wall come on as nearly the same course, as possible, except when the brick are not proportioned properly.

(h) Use only full headers (every brick a header) on the header courses of the interior tiers. See Bond Charts No. 8, No. 9, No. 10.

(i) If the wall is an uneven number of tiers thick, lay headers on the outside face tier, and on all filling tiers, then lay stretchers on the inside face tier. (See Bond Chart No. 5, No. 7, No. 10.)

On the next course above lay all headers on the inside face tier, and on the filling tiers, and then lay a stretcher on the outside face tier.

(j) On walls of uneven number of tiers thick, always lay the header on the outside face tier on the course below the header on the inside face tier, or the wall will have to be backed up solid, i. e., the entire width of the wall, on the inside to the height of the bottom of the header on the outside face tier before laying the header on the outside tier. This will make the inside face tier in the way of the bricklayer while he is laying the outside face tier.

(k) When a Flemish header is used on either the outside face tier or the inside face tier, always back up the stretcher of the Flemish header course with a stretcher, for if a header is put in this space it will upset the scheme of the bond as outlined above.

(l) When the top of a header occurring on either the outside face tier or the inside face tier is lower than the level of the filling tiers, level up on top of the header with a split brick and put on the headers exactly the same as herein described.

(m) There are countless other ways that a wall can be bonded so that it will serve its purpose, but the methods

here laid down are the quickest, strongest and most economical of time and money.

848

A careful examination of the racking ends of the isometric
figures will illustrate fully the actual working of the foregoing rules.

849

Where face brick do not lay up even courses, that is, even
heights, with the backing up brick, some arrangement must
be made whereby the backing is made level with the top of
the course that is under the header, to receive the face brick
headers.

850

There are three ways that this may be accomplished:

(a) By changing the number of courses between headers
on the face brick, and putting a header on every time the top
of the face brick comes level with the top of a "backing up"
course.

(b) By laying a course of split brick on the backing, to
come up level with the top of the face brick course that is
under the header.

(c) By laying a course of bull headers, that is, headers
laid on edge instead of in the usual way, on the flat. This
method should not be used, however, unless the brick are difficult
to split.

851

If the face brick header is a full header, this leveling
must be done the entire length of the wall. But if it is a
Flemish header, the leveling need be done only by a split bat
directly under, and a split bat over the Flemish header.

852

After the four inches directly back of the face four inches
are brought up to the level of the top of the course of face
brick directly under the header, the face brick header is laid.

853

Generally when a split is needed under the interior half
of the face brick header, another split is needed on top of the
interior half of the face brick headers to bring that tier (the
tier directly behind the face brick) back again to even courses
with the balance of the backing.

854

If this is required it should be done at once. Never bring the backing, instead, to the level of the top of the face brick header.

855

Common bond consists of several courses of stretchers followed by one row of either Flemish or full headers. See Bond Chart No. 1.

856

Use common bond everywhere on the inside face of all walls where strength, economy and speed only are desired.

857

Common bond is the strongest of all bonds. It is also the cheapest and quickest to lay. It should, therefore, be used on all walls over 8 ins. thick for backing up, regardless of the bond used on the face. It should be used for both sides of all walls where the greatest strength for the least cost is desired.

858

As to the number of courses of stretchers required between the headers, and as to whether the header course is to be of full headers or Flemish headers, depends on the quality of the bricks, their size, the kind of mortar used, the weight on the wall, the local building ordinance, and the architect's specifications.

859

Always use the strongest brick for the headers.

860

In the absence of orders or rules to the contrary, headers are to have not less than five courses of stretchers nor more than seven courses of stretchers between them.

861

The number should always be odd, so that the header course will start with the same bonds from the lead every time. The header course on the inside is to be put either on the same course or on the next above the header on the outside of the wall.

862

The headers are also to be laid in the middle of the wall with full header courses on the same course as the header on the line, regardless of whether the header on the line is a full header or a Flemish header.

863

The course above the header on the line is to be all full headers in the middle of the wall, breaking joints endwise on the headers on the course below.

864

Do not use a 2-in. piece instead of a ¾-in. closer in the header course to break the joints over the stretcher courses of common bond.

865

If the mortar has a large proportion of Portland cement and sand and the bricks are first class and very hard, the strongest bond is one Flemish header course to seven courses of stretchers.

866

If there is some doubt about either the quality of the mortar or the material of the headers, it should be a course of full headers to every seven courses of stretchers.

867

In countries subject to severe earthquakes, the best practice has been found to be five courses of stretchers followed by one course of full headers.

868

Fig. 164 shows a San Francisco building which was eight stories high, with self supporting brick walls. Common bond was here used, with five courses of stretchers between courses of full headers. This proves that every sixth course is as often as headers are needed, even in earthquake countries.

869

The generally accepted bond for the most economy and speed in the New England states is seven courses of stretchers between Flemish headers.

870

In no case should the header occur oftener than every sixth course, even in the most important and heaviest loaded work, for the reason that nearly every wall needs more stretchers than it has, and every course of headers deprives the wall of one course of stretchers on the outside face tier and on the inside face tier, and deprives the wall of stretchers on at least two courses on each interior tier.

871

The various tests made to determine the correct number

Fig. 164.—Wall of Mutual Life Insurance Co.'s Building, San Francisco, After the Fire.

of courses of stretchers between headers are misleading, because they have always been made on short lengths of walls which are not comparable with actual working conditions. When short lengths of walls are crushed, the oftener the headers occur, the higher the results; but under actual conditions, where long walls are built, the need of more stretchers due to uneven loading and settling, and shrinking is evidenced by the racking cracks that are not uncommon in all localities.

872

Many authorities claim that English bond and English cross bond, with the lap of only a quarter of a brick, are the strongest, because they have courses of full headers every second course. This is a great mistake, as is proved by the walls themselves. A wall failing for want of header courses is rare. Walls with racking cracks between the stretchers are very common.

873

Many old brick buildings show racking cracks on the face of the walls due to settlement or expansion and contraction due to changes in temperature, but few buildings can be seen with any damage or cracks due to splitting for want of headers.

874

We have taken down buildings with walls laid up 19 courses between headers and with no sign of splitting due to lack of headers.

875

We have never seen or heard of a case of a wall splitting lengthwise vertically with seven courses of stretchers between headers, and a careful examination of thousands of brick walls convinces us that the weakest point is the lack of stretchers, as shown by racking cracks, and that the more courses of stretchers, the more strength in this direction.

876

Superintendents will, however, use the bond called for in architect's specification.

877

Fig. 165 represents a photograph of the shops of the Leland Stanford, Jr., University at Palo Alto, California, taken immediately after the great earthquake of April 18, 1906. Care-

ful examination of the cracks of this building show the necessity of using more stretchers even at the sacrifice of headers.

878

Note that the brick is plastered to imitate stone work.

879

Figs 166 and 167 give additional evidence as to the number of courses of stretchers that should be laid between
header courses.

880

These figures show both sides of a 12-in. wall in a N. Y.
C. & H. R. R. R. freight shed. At the time of the accident the
wall was but a few days old, and the mortar still green. A

Fig. 165.—Three Earthquake Cracks, Showing Lack of Stretchers.

train of gravel cars was backed against the wall so hard as to
knock a hole clear through the wall. The wall showed no
injury at all, except as shown in the pictures.

881

The figures show how the wall looked immediately after
the accident, the only change being that the railway track
used for grading has been moved a few feet to the right.

882

The brick in this wall are laid common bond with seven
courses of stretchers between courses of full headers.

883

All other bonds than common bond are to be considered as purely ornamental.

Fig. 166.

Fig. 167.
Two Views of a Wall Pierced by a Gravel Train.

884

They can be divided into two classes:

(a) Those that change the bond at the corners, i. e., where all the even numbered courses are not plumb with one another.

(b) Those that do not change the bond at the corners, i. e., where all the even numbered courses are plumb with one another.

885

Those bonds that change the bond can be run out to the corner with a brick header 2 ins. wide on the corner. See Bond Chart No. 8, No. 14, No. 16. The bond of the corner can be maintained as headers and stretchers alternately and the bond can be changed by the usual practice of putting in two ¾ brick.

886

The bond is also sometimes maintained by putting a half brick near the corner; but this does not look so well. See Bond Chart No. 9, No. 10, No. 11.

887

The 2-in. piece next to an opening or a corner is not allowed in some specifications, but there are cases where it is absolutely necessary either to have the 2-in. piece, or else to have the bond out of plumb.

888

The 2-in. piece is sometimes placed one-half brick from the corner, see Bond Chart No. 5, and sometimes put on the corner such as a header 2 ins. wide. See Bond Chart No. 12, No. 13.

889

The position of the 2-in piece is determined by the following features:

(a) Whether or not the pattern of the bond is such that the pattern runs around the corner. See Bond Chart No. 14.

(b) Whether or not a border effect on the corner is desired. See Bond Chart No. 9, No. 10.

(c) Whether or not the location of the 2-in. piece interferes with the symmetry of the pattern. See Bond Chart No. 15.

890

In the absence of any definite instructions from the architect, ornamental bonds are to have at least one header to every ten stretchers showing on the face of the wall, and, if the bond will not permit as many headers as this showing on the face, then additional blind or diagonal and galvanized iron headers must be put in at the rate of one additional blind header for every five extra stretchers that show on the face of the wall.

891

"All-Stretcher Bond" is the correct name for the arrangement of brick that shows nothing but stretchers on the face of the wall. See Bond Chart No. 2.

892

This bond is also sometimes called "running bond," "plumb bond" or "stretcher bond."

893

The building ordinances of the city of New York say that walls of all stretcher bond "shall be 4 ins. thicker than the walls are required to be under any section of this code."

894

This bond is the weakest of all, unless it is tied in with brick that are 8 ins. square. Brick of this shape are hard to obtain in most localities, and are generally of a slightly different shade or thickness from the standard brick.

895

If the bonding stretchers $(8 \times 8 \times 2\frac{1}{2})$ are not used, diagonal headers must be used sometimes. The latter are not satisfactory, as is shown by Fig. 163, and should be supplemented by metal ties.

896

If metal ties are used extreme care should be exerted to have them completely covered with Portland cement or they will surely rust out sooner or later, even if galvanized.

897

Many walls can be seen in the old South Cove section of Boston which have split from the top to bottom, due to the rusting off of metal ties.

898

Many such walls may be seen there, also, which have been repaired by cutting holes completely through the walls two feet long, two courses high and five feet apart on centers horizontally, and seven courses between vertically, and filling these holes completely full of headers.

899

Walls repaired in this manner have become serviceable again without further attention or repairs.

900

Flemish bond consists of courses of alternate headers and stretchers, with the center of the headers located plumb over the center of the stretchers beneath them.

901

Under no consideration must two stretchers adjoin each other in the same course.

902

No joint must come under or over a header even in the smallest pier, return or reveal. This is a matter often disregarded but must never occur on our work.

903

For the strongest, quickest and most economical results, real headers should not be extended into the backing oftener than every sixth or eighth course. The headers in the five or seven intervening courses should be imitation, or bat, headers.

904

There are several ways of laying out the bond on the leads.

905

The bond on each face from the corner, and also out from the angle, can be a header with a three-quarter over and under it (See Bond Chart No. 4), or it can be made with a header and 2-in. piece with a stretcher over and under it (See Bond Chart No. 5), or it can be made with the end of the brick made 2 ins. wide with a half brick over and under it (See Bond Chart No. 3), or any combination of these three lay-outs.

906

This last method does not look as well as that of Bond Chart No. 4, No. 5, and should be used only when the lay-out fits the short piece of wall that way.

907

It is perfectly good practice to use, or not to use, the 2-in. piece near the corner and near the angle, but, as it gives a border effect to the end of the wall, if used at all it should be used on both ends of the wall and on each side of the corner and the angle, if possible, for the sake of symmetry.

908

When the 2-in. piece is used next to the header on a corner, it should always be considered a part of the stretcher that goes along side of it. On black header or other colored pattern Flemish bond, the 2-in. piece must be colored the same as the stretchers and must never be considered as a half header.

909

In running into an angle the end of a stretcher sometimes comes to within 2 ins., or half a header of the angle. In this case, of course, the exposed 2 ins. between the stretcher and the angle is half a header, and must be colored as a header.

910

Take extra care to have the two end joints the same width on the 2-in. piece. If care is not taken on any cut brick the cut side of the brick will make the narrower joint of the two.

911

Flemish bond has no decided diagonal lines, and consequently it is not so important to have the courses of brick laid to exact heights as it is that the horizontal joints be the same thickness for uniform appearing work.

912

It is essential that the plumb bond pole be used, or the inaccuracy will be very noticeable.

913

Flemish bond is much cheaper to lay than English bond or English cross bond, when the brick are not proportioned on the basis of the length of two headers plus one joint equaling the length of one stretcher.

914

If the brick are proportioned as above, the cost of the three above mentioned bonds is about the same.

915

English bond consists of alternate courses of full headers and stretchers, the headers being plumb over each other and the stretchers being plumb over each other, the headers being divided equally over the stretchers and over the joints between the stretchers. See Bond Chart No. 6, No. 7.

916

The bond can start from the angle and from the corner after the rules of Flemish bond.

917

If bricks of different colors are used with English bond, it is desirable that the pattern start the same distance from each side of the corner and from each side of the angle.

918

It is also generally desirable that the header comes on the

same course on all of the walls, angles, and around the corners, all the way around the building, for the reason that, owing to the extra number of joints, the header course shows up a different shade of color and presents a belt-like appearance.

919

If the brick are not made so that the length of two headers plus one joint is equal to the length of one stretcher, this bond will require a great deal of cutting and placing of occasional long bats in the header course or of clipping the ends off all stretchers or at regular intervals.

920

The plumb bond pole is necessary for the best results. With English bond, when bricks vary in thickness, laying to uniform heights for each course is not so important as having the same thickness of horizontal joints.

921

If English bond or any of the cross bonds are to be used, due precautions in selection must be taken at the time that the brick are purchased to insure getting brick that are correctly proportioned.

922

The majority of American brick do not lend themselves readily to any of the "cross bonds" because they are generally made with too great a length in proportion to their width. The result is that two headers plus one end joint is not as long as one stretcher.

923

English cross bond is the same as English bond, except that the stretcher courses break joints evenly over each other. See Bond Chart No. 8, No. 9, No. 10.

924

This feature will either change the bond every course on the corner or next to the corner.

925

It can be changed next to the corner by putting in a half brick, or by two three-quarter bricks in every second stretcher course on each side of the corner and each side of the angle.

926

Which of the above three methods should be used depends on several conditions, as follows:

(a) Whether a border effect is desired on the corner. See Bond Chart No. 9, No. 10.

(b) Whether the headers are desired on the same course for a continuous bond effect, as in Bond Chart No. 9, No. 11, or on different courses, as in Bond Chart No. 8, No. 10.

(c) Whether the bond is run out to the corner and to the angle using the 2-in. end on the header end of the corner brick. See Bond Chart No. 8, No. 11.

(d) Whether it is desired to have the diagonal lines run around the corner in an unbroken line. See Bond Chart No. 8, No. 11.

927

English cross bond lends itself particularly well to ornamental brickwork on account of its diagonal lines of joints and the St. George's cross effect. See Bond Chart No. 26 and No. 23.

928

For the best effects it is necessary to use the plumb bond pole and the story pole with the bricks laid to exactly even heights, or the diagonal lines will not be straight.

929

For best results it is more important that bricks are proportioned correctly than for English bond.

930

Garden wall bond consists of three stretchers to one header on every course. See Bond Chart No. 13.

931

Garden wall bond is especially adapted to boundary walls and any other walls not over 8 ins., or two brick, thick.

932

As each header is completely surrounded by stretchers, the bricklayer can favor the inaccuracies in length of the headers by averaging the position of the faces of the adjoining stretchers in and out from the ashlar line.

933

Flemish spiral bond is particularly adapted to chimneys and stairway towers and bays of small radius, as it cuts down the amount of rolling necessary on curved surfaces. See bond chart No. 72.

934

The bond can be swung back and forth if desired, or carried all in one direction. See bond chart No. 73.

935

The general appearance of any brickwork where each course ends with a cut brick, such as paved work, will be much improved if the piece is placed so that the cut edge of the closer is away from the end of the work.

936

This hides the irregularities of the cut edge.

937

If the cut edge is placed on the exterior edge of the paving, or against the border of the paving, the inaccuracies of the cut edges will be exaggerated; first, because the cut edges appear where the eye unconsciously expects them; and, second, because they are all together in a straight line.

938

On herringbone paving, and in fact all other patterns of paving, do not cut the pieces and closers until the large portions of the body of the paving are done.

939

An analysis of motion study will instantly reveal a tremendous saving of motions and time, if the closers are all made and placed at one operation.

940

When paving, whether in sand or mortar, prepare guides for a template or straight edge to slide upon that will spread off the bed evenly to receive the brick with the least amount of tamping.

941

In other words, arrange the beds with a guide so as to require just enough tamping to bed the brick properly.

942

Ornamental bonds should not be made to appear as complicated and mysterious as they were in the days when arch cutters housed themselves in shacks to prevent others seeing the method used to cut bricks.

943

The place that requires the services of the most skilled bricklayer is in reality a perfectly plain blank wall, where the slightest deviation from accuracy and uniformity will show up with the most prominence.

944

A steel tape is often used by the foremen to mark out joints in a wall containing ornamental brickwork. This method of marking should not be used as it delays the brick-layers too long.

945

The foreman should provide bricklayers with plumb bond poles with joint notches cut in each of the four edges. See Fig. 110. The poles should have marked near each edge the letter or symbol of the courses to which that edge applies.

946

A plumb bond pole thus marked, in combination with a story pole, will take care of all racking patterns. They will require no other attention, no matter how long the rack is.

947

All marks on the plumb bond pole for making plumb bond should be made at the center of the vertical joint, not at the edge of the brick, except where wide vertical joints are used. See Fig. 110.

948

Marks on the story pole should represent the top edge of the brick, not the center of the mortar joint.

949

This method will simplify and make accurate those problems of bricklaying that have always been considered difficult by bricklayers and foremen. It will reduce delays, conferences and arguments on the scaffold. It will put all of the laying out in the hands of the foreman, who can get the wall laid out exactly as he lays it out on a pole beforehand.

950

Not only must the brick be laid with plumb bond, and with courses at even heights, but they must be also all of one cull. If they are not, they should not be culled at all, or the difference in thickness of the horizontal and also vertical joints will be very noticeable, especially if the wall has a decidedly diagonal pattern on the face.

951

When the desired effect of bond is obtained by different widths of the vertical joints, as in bond charts Nos. 14, 15, 17, 61, instead of different colors or culls of brick, the plumb bond pole must be notched with one notch at the center of each narrow joint and two notches at each edge of each wide vertical joint.

952

On large patterns of bond, as for example the Flemish double cross bonds, it is sometimes quite impracticable to have the work symmetrical on all corners. In such cases the foreman must start the pattern at the most important corner, or the corner most seen, or else divide the remaining portions of the patterns evenly at each end of the wall. See bond chart No. 15, No. 16.

953

In case the bond does not work out symmetrically at all corners, judgment must be used to decide which wall it is most important to have symmetrical at both ends.

954

Ornamental patterns in brickwork are expensive, if much measuring and cutting of brick to length is required.

955

Almost any decorative pattern can be built with little or no cutting, if the work is laid out with the right combinations of courses of stretcher bond, English cross bond, Flemish cross and Flemish double cross bonds.

956

The advantage of using the particular bond that is best adapted to the particular pattern, is due to the fact that fewer joints are used by some bonds than others in those places where joints are not required.

957

These joints that are not needed can be made much narrower than the others, thus being scarcely noticeable. See bond chart No. 54.

958

When ornamental bond and patterns are designed without a definite drawing showing clearly each and every brick, the foreman should lay out on the standard scale paper a combination of brick that will show the fewest number of joints that do not form a necessary part of the pattern.

959

The diagrams shown in this system are given by way of assistance to the foremen in laying out bond to fit the architect's design. It is not expected that they will exactly fit, but that they will show the key to methods that will tend to reduce costs by obviating some measuring and cutting.

960

Following the methods shown here will surely result in cutting down the labor on decorative work.

961

Consult the office drawings for additional keys for special cases.

962

After determining the particular combination of bonds that will carry out the architect's design with the lowest labor cost, submit the same for the architect's approval before proceeding with the work.

963

A very large proportion of the ornamental bonds can be reduced to a simple bond of symmetrical arrangement of headers and stretchers by arranging certain units with or without borders.

964

Each one of these units shown, bond chart No. 25, varies two courses in height and one-half brick in width.

965

These units can be made by various arrangement of headers and stretchers, but the patterns of the units as drawn are the least expensive from a labor standpoint.

966

They will have a definite symmetrical bond easily remembered by the bricklayer without constantly looking at a detail drawing, and they will carry out any pattern with the least amount of cutting of brick to odd or uneven lengths.

967

These patterns are the most economical where the brick are of such proportions that two headers plus one joint do not exactly equal one stretcher in length. Furthermore, they require the least possible number of pieces to lay to the line.

968

This last point is a great factor in reducing labor costs, as very few brickyards in America make any attempt to produce brick of exactly correct proportions.

969

Those brick that are supposed to be of the correct relative measurements are not always so, because of unequal shrinkage while baking.

970

It will be noticed that these units are 1, 2, 3 and 4, alone, or with borders around them.

971

Some of these ornamental patterns can be executed in color with English cross bond instead of the bonds as shown here, but it would not only be much more expensive but would not be nearly so accurate, due to the difficulty of keeping the header course from running ahead or behind the bond and the consequent difficulty of making the joints in the header course uniform and the same size as those in the stretcher courses.

972

There are countless other arrangements of joints than can be made to carry out these same patterns, but the arrangements as shown, made of the standard "bond units," will require fewer standard courses and consequently fewer plumb bond poles than any other arrangement. All of which, consequently, cuts down the complications of the bricklayer's work, reducing cost and increasing speed.

973

A large number of examples of regular bond are given here, to show that any pattern or bond can be best handled by the foreman and the bricklayer if he is thoroughly conversant with the laws governing bonds.

974

By using the methods here shown any unusual bond, if regular, that is, if it repeats itself, can be laid accurately with no other precautions than a properly laid out plumb bond pole and story pole.

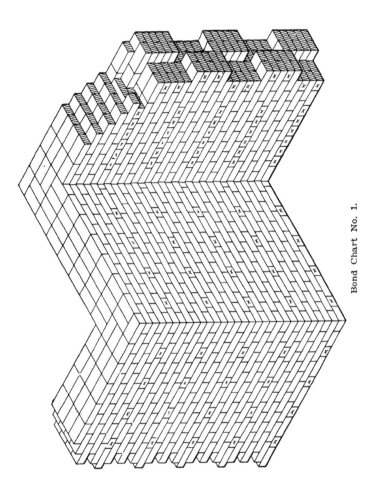

Bond Chart No. 1.

Bond Chart No. 1. Isometric view showing common bond with five courses of stretchers between header courses.

The walls to the left and to the center show courses of Flemish headers, while the wall to the right shows courses of full headers.

The corners should start with full length headers and stretchers every course except the header courses, which should be either a three-quarter brick or a half brick plus a three-quarter brick from the corner.

On short lengths of wall it is good practice to run the bond straight into the angle as it works out from using full length brick. On long lengths of wall it is generally better practice to start the angles exactly the same as described for the corner, and to have the piece occur in the middle of the course between leads.

This bond with the header course made with flemish headers is the strongest and most economical of all the bonds. On certain work it is less objectionable than a bond having full header courses. Courses of full headers always present a banded effect, owing to a much larger number of joints, and, on culled work, owing to the end of a brick being generally a slightly different color than the side of a brick.

The blocking on the end of this wall shows the right way to build blocking on common work.

The headers should never be put nearer to the blocking than as shown. The purpose of the blocking is to make a tie lengthwise of the wall and joints, and additional headers would tend to make a weakness at this point.

The headers that are bound to occur in the end of the blocking more than compensate for the loss of headers in the header course. A cross on a brick denotes a header.

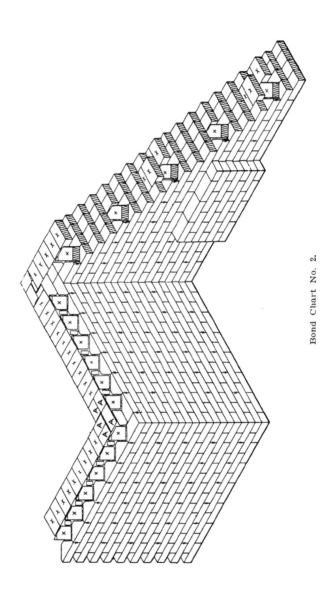

Bond Chart No. 2.

Bond Chart No. 2. Isometric view showing all-stretcher bond on the exterior face tier, with blind headers wherever there is a cross. The interior face tier is common bond.

All-stretcher bond should always start from the corner with a full, length header and a full length stretcher. The piece, if any, should be placed either in the middle of the course between the two leads, or adjoining the angle.

It is always desirable to start the angle with a full length header and stretcher, but it costs less to reduce the length of the half brick and stretcher adjoining the angle, if that will save cutting between the leads.

This bond should never be used except under orders from the architect, as it is more expensive and not so strong as any of the bonds that have real headers.

The pilaster shows that any projection of whole brick built from a wall does not change the bond from what it would be if it ran straight through and past the pilaster.

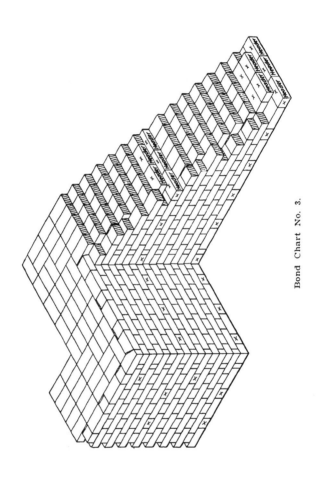

Bond Chart No. 3.

Bond Chart No. 3. Isometric view showing Flemish bond on the exterior face tier and common bond on the interior face tier.

The bonds to the left of the corner and to the right of the angle are alike, and the bonds to the right of the corner and to the left of the angle are alike.

It is generally desirable, when one of the faces running to the angle shows a 2″ piece or half a header adjoining an angle, to put a 2″ piece on the other wall also next to the corner instead of as shown to the right of the angle in this figure.

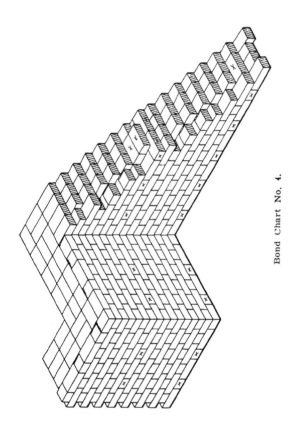

Bond Chart No. 4.

Bond Chart No. 4 Isometric view showing Flemish bond on the exterior face tier and common bond on the interior face tier.

The bond on both sides of the corner and angle is symmetrical, starting with a half brick and a three-quarter brick alternately on each side of the corner and angle.

The 2″ piece is permissible at any time that the length of the wall spaces out for the 2″ piece, but it is generally better to make the corner or the angle look as symmetrical as possible, if it can be done without additional cutting.

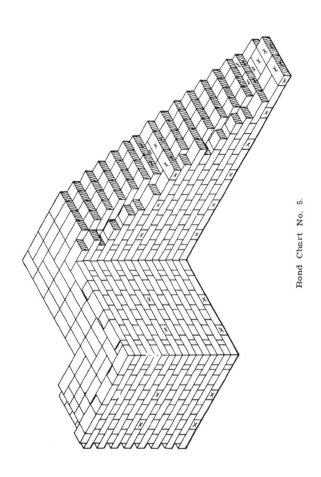

Bond Chart No. 5.

Bond Chart No. 5. Isometric view showing Flemish bond on the exterior face tier and common bond on the interior face tier.

The bond on both sides of the corner and angle is symmetrical, the header starting with a whole brick, and a half brick plus one-quarter brick plus a whole brick, alternately, on each side of the angle and the corner.

In Flemish bond the 2″ piece must never occur between two headers nor over a joint, and should be used only against the brick that comes next the corner or angle.

In Flemish bond the header must never come over or under a joint and must always be plumb over the center of the stretcher under it and over it.

All odd numbers of courses must be alike with joints plumb over each other and all even numbered courses must be alike with joints plumb over each other.

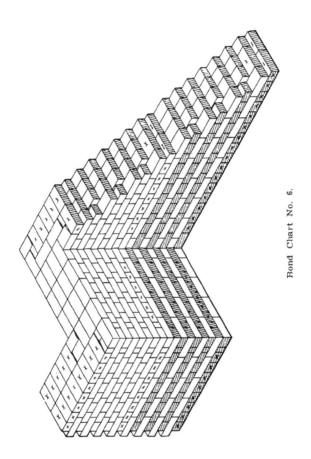

Bond Chart No. 6.

Bond Chart No. 6. Isometric view showing English bond on the exterior face and common bond on the interior face tier.

This bond is very easy to lay out, but is quite different when headers are on the same course on both sides of angles and corners than from where they are on alternating courses.

This figure shows both faces of the angles symmetrical, the headers being on different courses wherever the face of two walls intersect.

This is the easiest way to lay out English bond, but it is often undesirable as the headers do not occur on the same courses.

Full headers should not be used oftener than every sixth or eight course.

We realize that this is contrary to much that has been written, but it is undoubtedly in the interests of the best work, as is shown by countless examples of old work laid English bond, where racking cracks invariably show themselves before there is any splitting between tiers when laid with less than seven courses between headers.

This is so contrary to present beliefs and practice, that we desire to call particular attention to this fact.

The header starts on each side of the corner a half brick and a 2″ piece from the corner.

The header starts from each side of the angle three-quarters of a brick from the angle.

If desired, a 2″ piece and a header may be used in place of the three-quarter brick at the angle, but it makes many more joints in the angle and is not so desirable.

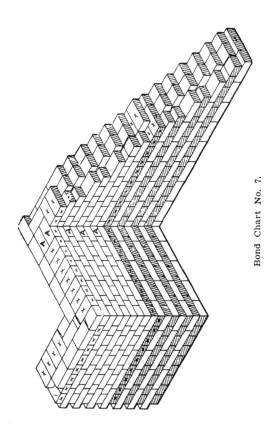

Bond Chart No. 7.

Bond Chart No. 7. Isometric view showing English bond on the exterior face tier and common bond on the interior face tier.

The headers on the left section are on a different course from the headers on the central and right section.

Headers should always occur on the same course all the way around the building or else change at every corner and angle.

There are many ways that English bond can be laid from the corner and the angle.

The rules to be followed are: (a) no lap shall ever be less than one-quarter of a brick; (b) the header must be divided evenly over each joint, or over the center of each brick.

Corresponding joints in different courses must be plumb over each other, every header course being like every other header course and every stretcher course like every other stretcher course.

Bricks marked "A" may be either stretchers or headers.

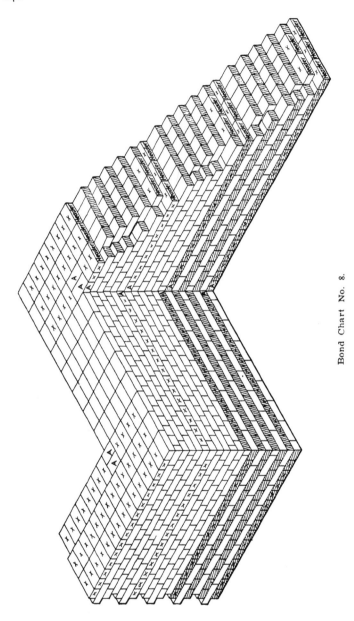

Bond Chart No. 8. Isometric view showing English cross bond on the exterior face tier and common bond on the interior face tier.

The header courses occur on different courses on adjoining walls.

The change in the bond is made at the corner.

The diagonal lines run upward in the center wall and are continued around the corner in the wall to the left, but the diagonal lines in the wall to the left are not continued around the corner in the center wall.

The header occurs a half brick from the corner to the left of the corner and a quarter brick from the corner to the right of the corner.

The bond changes at the angle by continuing the stretcher courses into the angle without cutting.

The header occurs one-quarter brick to the left of the angle and a half brick from the right of the angle.

The diagonal lines extend up the center wall around the angle and upward on the wall to the right, but the diagonal lines on the wall to the right do not continue past the angle upwardly in the center wall.

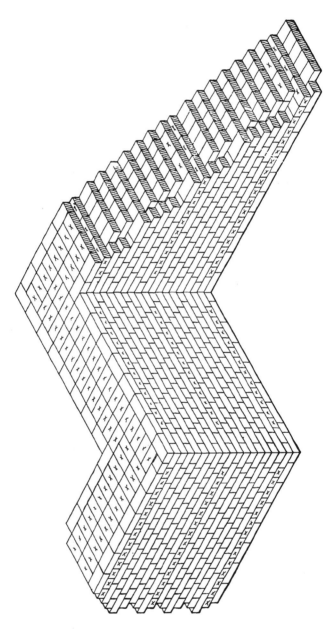

Bond Chart No. 9.

Bond Chart No. 9. Isometric view showing English cross bond on the exterior face tier and common bond on the interior face tier.

The header courses occur on the same courses on all walls.

The corner is built without changing the bond at the corner.

The change occurs every fourth course by inserting a half brick at the distance of a whole brick from the corner at the right of the corner, and a half brick from the corner to the left of the corner.

The bond on each side of the angle is symmetrical, the bond changing on each side of the angle every course.

The header course starts three-quarters of a brick from the angle on each wall.

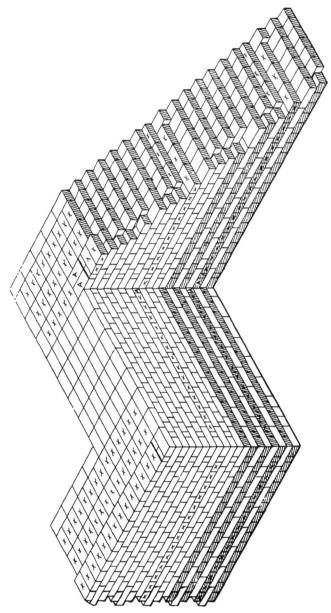

Bond Chart No. 10. Isometric view showing English cross bond on the exterior face tier and common bond on the interior face tier.

The header courses occur on different courses on adjoining walls.

The corner is built without changing the bond at the corner.

The change occurs by inserting a half brick every fourth course at the distance of a whole brick to the left and also to the right of the corner.

The two faces of the angle are made differently, but could be made exactly alike by transposing the 2″ piece with the header in every header course, and by transposing the header and the whole brick in every second stretcher course on either side of the angle.

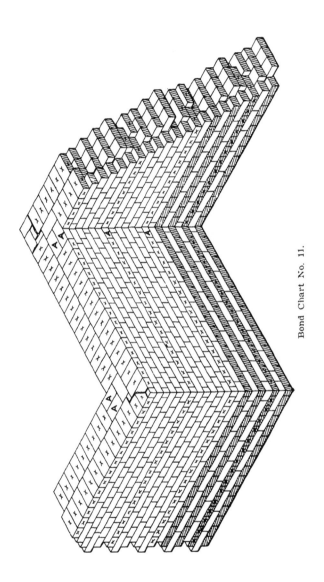

Bond Chart No. 11.

Bond Chart No. 11. Isometric view showing English cross bond on the exterior face tier and common bond on the interior face tier.

The header courses occur on the same course on all walls.

The corner is built alike on both sides. The diagonal line of joints runs unbroken on one wall around the corner and on the other wall.

This bond is particularly desirable where lines of diagonal joints are desired, but the corner itself does not look as massive and strong as it would if the brick showing 2″ on each face were made as a ¾ on one face.

The bond is changed every course at the corner.

The header occurs 2″ from the corner on each face.

The angle is symmetrical on both sides. The bond is changed every course at the angle by running out the stretcher courses into the angle as they occur.

This maintains the diagonal lines up one face, around the angle and on the other face.

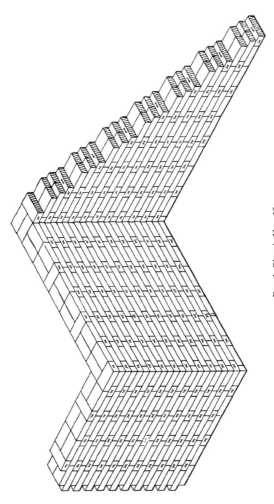

Bond Chart No. 12.

Bond Chart No. 12. Isometric view showing header two-stretcher garden wall bond.

When this bond is used on garden or boundary walls, it is good practice to have every header a real header extending through the wall, showing the same pattern on both the exterior and interior face tiers.

When this pattern is used on walls thicker than 8″, it is generally advisable to have real headers only on every fourth course.

The corners and angles may be laid as shown, or in accordance with any of the rules for laying corners and angles in English bond.

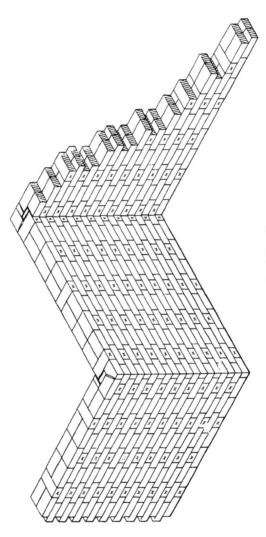

Bond Chart No. 13.

Bond Chart No. 13. Isometric view showing header three- stretcher garden wall bond.

When this bond is used on garden or boundary walls it is good practice to have every header a real header extending through the wall, showing the same pattern on both the exterior and interior face tiers.

When this pattern is used on walls thicker than 8″, it is generally advisable to have real headers only on every second header course.

Walls exposed on both sides to frost and weather are more apt to split between the tiers than walls used in building construction, and consequently the headers should occur oftener.

The corners and angles may be laid as shown, or in accordance with any of the rules for laying corners and angles in Flemish and English bond.

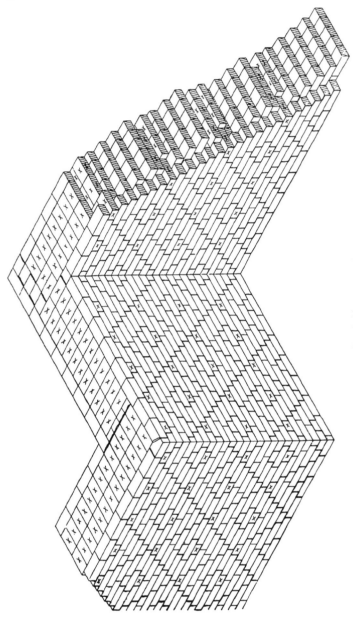

Bond Chart No. 14. Isometric view showing Unit 6 on the exterior face tier, divided symmetrically on the corner and evenly on the angle.

The interior face tier is common bond.

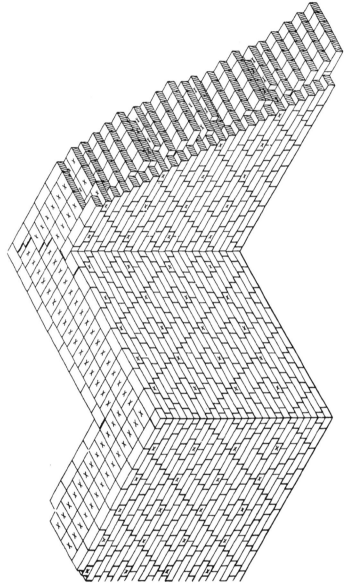

Bond Chart No. 15. Isometric view showing Unit 6 on the exterior face tier, divided evenly on the corner and symmetrically on the angle.

The interior face tier is common bond.

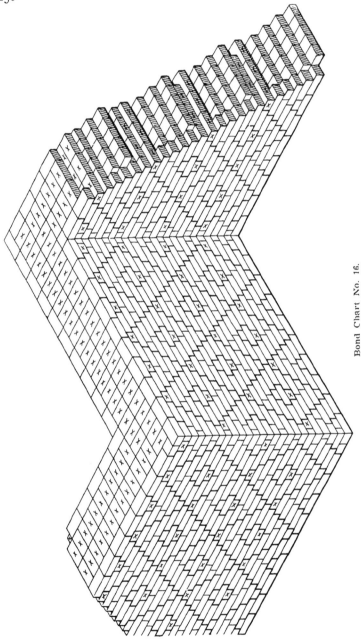

Bond Chart No. 16.

Bond Chart No. 16. Isometric view showing Unit 6 **on** the exterior face tier, divided symmetrically on the corner and on the angle.

The interior face tier is common bond.

On the bond charts of ornamental bond which follow, charts No. 17 to No. 71, the vertical arrangement of the headers is shown by the vertical key, which is hatched in. The horizontal arrangement of the headers is given in the captions.

Bond Chart No. 17.

Bond Chart No. 17 is a Flemish cross-bond where the header is shifted its own width back and forth from a vertical line.

All odd numbered courses are Flemish header courses, and all even courses are stretchers crossed every stretcher course.

The different shading shows effects that can be obtained by different culls of brick or by wide joints.

Bond Chart No. 18.

Bond Chart No. 18. The bond of Bond Chart 18 is the same as Bond Chart 17, but the cross-hatching is located differently.

Bond Chart No. 19.

Bond Chart No. 19. This bond is like Flemish bond, except that the Flemish header courses are separated by two courses of quarter lapped stretchers; the laps travel two courses to the left and then two courses to the right alternately.

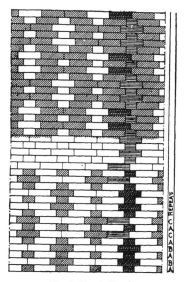

Bond Chart No. 20.

Bond Chart No. 20 is Flemish cross bond. The odd courses are Flemish bond, and the even courses are stretchers crossed every second time.

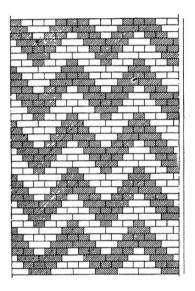

Bond Chart No. 21.

Bond Chart No. 21 is an English cross bond, a pattern being brought out by the use of two culls of brick.

Bond Chart No. 22.

Bond Chart No. 22 and Bond Chart No. 23 are the same pattern laid out with header two-stretcher cross-bond and English cross-bond, the only difference being in the number of joints which is much larger in No. 23, the English cross-bond, than in No. 22.

Many of these patterns for which special bonds have been shown can be executed in English cross-bond, but the English cross-bond will require more labor, especially if the bricks are not proportioned so that two-headers plus one joint equals the length of one stretcher.

Bond Chart No. 23.

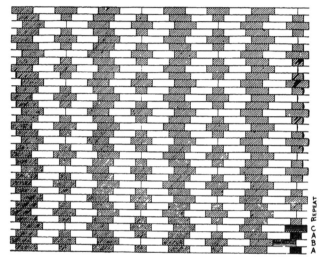

Bond Chart No. 24.

Bond Chart No. 24 is a garden wall cross bond, the odd courses are header three stretchers, and the even courses are stretchers crossed.

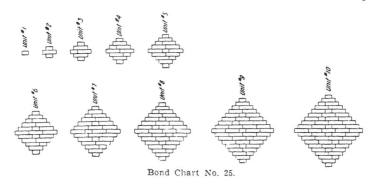

Bond Chart No. 25.

Bond Chart No. 25 shows ten units. Each successive one is a half brick wider and two courses higher than the one that precedes it.

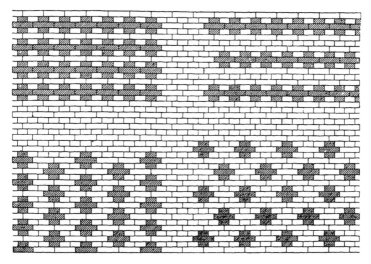

Bond Chart No. 26.

Bond Chart No. 26 is unit 2, shown (a) by itself, (b) with a horizontal border, (c) with a vertical border, and (d) with a horizontal and a vertical border. It consists of English cross-bond.

Unit No. 2 is the foundation of English cross bond, English bond and Flemish bond.

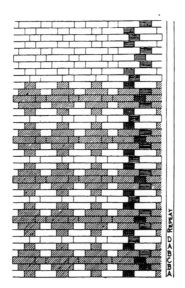

Bond Chart No. 27.

Bond Chart No. 27 is unit 3. It consists of Flemish bond in every course.

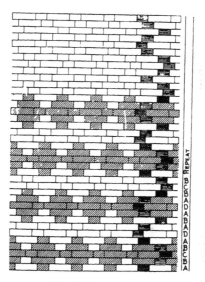

Bond Chart No. 28.

Bond Chart No. 28 is unit 3 with a horizontal border. It consists of Flemish bond in every course.

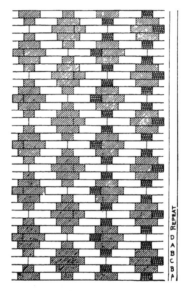

Bond Chart No. 29.

Bond Chart No. 29 is unit 3, with a vertical stretcher border. It consists of header, three stretchers in every course.

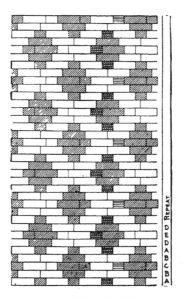

Bond Chart No. 30.

Bond Chart No. 30 is unit No. 3, with a horizontal and a vertical stretcher border. It consists of Flemish semi-cross bond. The odd numbered courses being Flemish headers, and the even numbered courses being stretchers crossed on each other every second time.

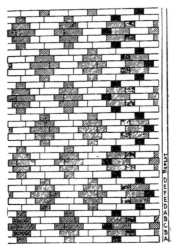

Bond Chart No. 31.

Bond Chart No. 31 is unit No. 3 with a horizontal and a vertical stretcher border. It consists of header two stretchers on every course. This resembles No. 30, but has a straight border.

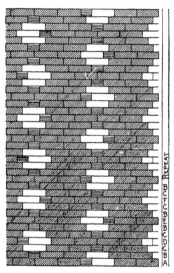

Bond Chart No. 32.

Bond Chart No. 32 is unit No. 3 with a two stretcher vertical border. It consists of header five stretchers on every course.

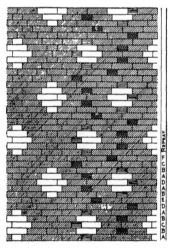

Bond Chart No. 33.

Bond Chart 33 is unit No. 3, with a horizontal and a vertical double stretcher border. It consists of header, three stretchers in every course.

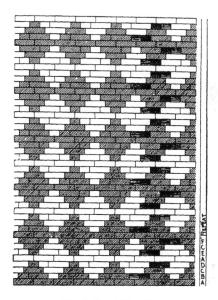

Bond Chart No. 34.

Bond Chart No. 34 consists of unit No. 4 with the odd numbered courses consisting of crossed stretchers, and the even numbered courses consisting of two-headers, stretcher.

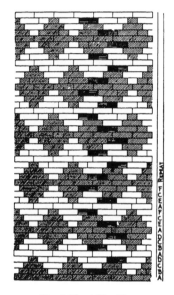

Bond Chart No. 35.

Bond Chart No. 35 consists of unit No. 4 with a horizontal stretcher border.

All odd numbered courses are crossed stretchers and the even numbered courses are two-headers, stretcher.

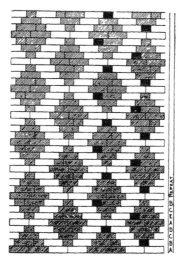

Bond Chart No. 36.

Bond Chart No. 36 consists of unit No. 4 with a vertical stretcher border, odd numbered courses being crossed stretchers, even numbered courses being header, one-stretcher, header, two-stretchers.

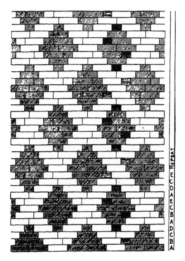

Bond Chart No. 37.

Bond Chart No. 37 consists of unit No. 4 with a horizontal and a vertical stretcher border, the odd courses being crossed stretchers, the even courses being Flemish headers.

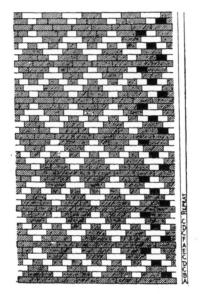

Bond Chart No. 38.

Bond Chart No. 38 is unit No. 4 with a horizontal header border.

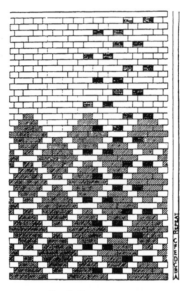

Bond Chart No. 39.

Bond Chart No. 39 is unit No. 4 with a horizontal and a vertical header border.

Bond Chart No. 40.

Bond Chart No. 40 is unit No. 4 with a wide horizontal border. It consists of two headers stretchers crossed on every even course, and stretcher on every odd course.

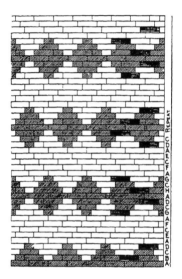

Bond Chart No. 41.

Bond Chart No. 41 is unit No. 4 with a horizontal border five stretchers wide. It consists of alternating courses of crossed stretchers, the courses between being two headers, stretcher. The border is of quarter lapped courses of stretchers, the laps changing their direction at the units.

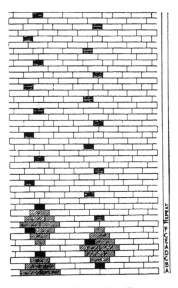

Bond Chart No. 42.

Bond Chart No. 42 is unit No. 4 with a two stretcher vertical border. It consists of crossed stretchers in odd course, and of header two stretchers, header three stretchers in every even course.

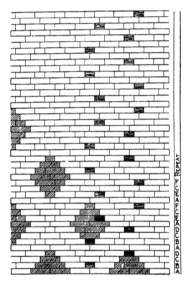

Bond Chart No. 43.

Bond Chart No. 43 is unit No. 4 with a two stretcher horizontal and a two-stretcher vertical border. It consists of crossed stretchers in every odd course, and of header stretcher, header two-stretchers in every even course.

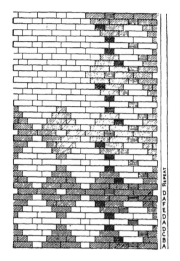

Bond Chart No. 44.

Bond Chart No. 44 is unit No. 5. It consists of header two stretchers in every course.

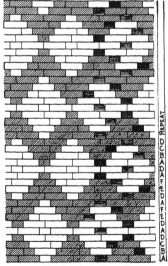

Bond Chart No. 45.

Bond Chart No. 45 is unit No. 5 with a horizontal stretcher border. It consists of header two stretchers in every course.

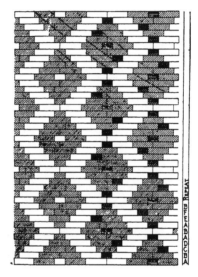

Bond Chart No. 46.

Bond Chart No. 46 is unit No. 5 with a vertical stretcher border. It consists of header four stretchers on every course.

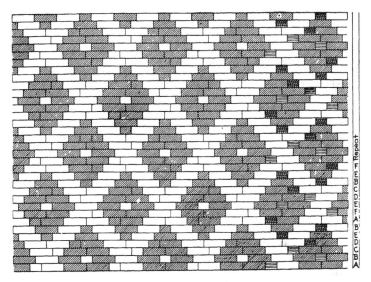

Bond Chart No. 47.

Bond Chart No. 47 is unit No. 5 with a horizontal and a vertical stretcher border.

This consists of header three-stretcher in every course.

Bond Chart No. 48.

Bond Chart No. 48 is unit No. 5 with a double-stretcher horizontal border. Each course consists of headers two-stretchers.

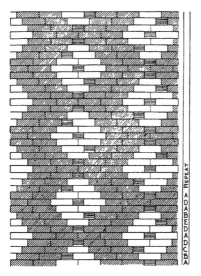

Bond Chart No. 49.

Bond Chart No. 49 is unit No. 5 with a vertical double-stretcher border; it consists of header six-stretchers in every course.

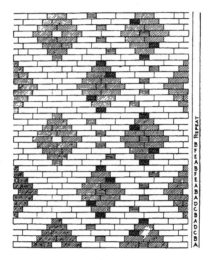

Bond Chart No. 50.

Bond Chart No. 50 is unit No. 5, with horizontal and a vertical double stretcher border. It consists of header, four stretchers in every course.

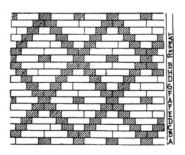

Bond Chart No. 51.

Bond Chart No. 51 is unit No. 5 with a horizontal and a vertical header border.

Bond Chart No. 52.

Bond Chart No. 52 is unit No. 5 with a header two-stretcher header horizontal and vertical border.

Bond Chart No. 53.

Bond Chart No. 53 is unit No. 6. This is a Flemish header cross bond. It consists of stretchers crossed on every odd course, and of Flemish bond on every even course.

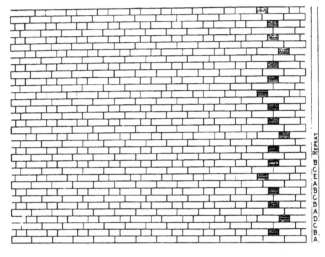

Bond Chart No. 54.

Bond Chart No. 54 is unit No. 6 with some joints exaggerated and some joints diminished.

This is a Flemish header cross-bond, the odd courses being crossed stretchers, and the even courses being Flemish headers.

Bond Chart No. 55.

Bond Chart No. 55 is unit No. 6 with a horizontal stretcher border. It consists of Flemish courses every odd course, and of stretchers crossed every even course.

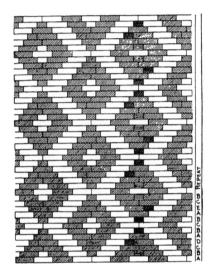

Bond Chart No. 56.

Bond Chart No. 56 is unit No. 6, with a vertical stretcher border. It consists of header, two stretchers, cross bond. The odd numbered courses are stretchers crossed on each other every course, and the even numbered courses are—header, two stretchers.

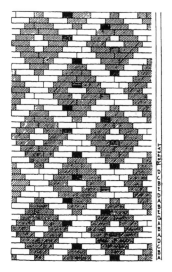

Bond Chart No. 57.

Bond Chart No. 57 is unit No. 6, with a horizontal and a vertical stretcher border. It consists of header, one stretcher; header, two stretchers, cross bond,—that is—the odd numbered courses are header, one stretcher, and header, two stretchers,—and the even numbered courses are—stretchers crossed on each other every course.

Bond Chart No. 58.

Bond Chart No. 58 is unit No. 6 with a horizontal two-stretcher border. It consists of stretchers crossed on the odd courses and Flemish bond on the even courses.

Bond Chart No. 59.

Bond Chart No. 59 is unit No. 6 with a vertical two-stretchers border. It consists of stretchers crossed on the odd courses and header three stretchers on the even courses.

Bond Chart No. 60.

Bond Chart No. 60 is unit No. 6 with header two stretcher header border horizontally and vertically.

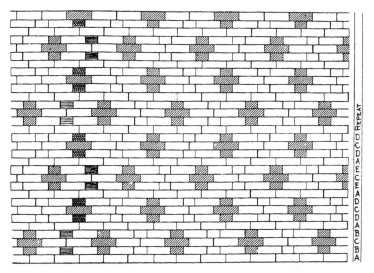

Bond Chart No. 61.

Bond Chart No. 61 bears a close resemblance to Bond Chart No. 54. It consists of stretchers crossed on every odd course, and of Flemish courses every even course.

Bond Chart No. 62.

Bond Chart No. 62 is the same as unit No. 6, except that the brick of the stretcher course is divided evenly on the header of the course over it instead of being crossed every second stretcher course.

This makes the stretchers cross twice to every three stretcher courses.

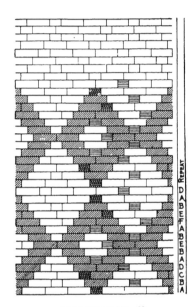

Bond Chart No. 63.

Bond Chart No. 63 is unit No. 7. It consists cf header three stretchers in every course.

Bond Chart No. 64.

Bond Chart No. 64 is unit No. 7 with horizontal stretcher border. It consists of header, three stretchers in every course.

Bond Chart No. 65.

Bond Chart No. 65 is unit No. 7 with a vertical stretcher border. It consists of header five stretchers in every course.

Bond Chart No. 66.

Bond Chart No. 66 is unit No. 7 with horizontal and a vertical stretcher border. It consists of header, four stretchers in every course.

Bond Chart No. 67.

Bond Chart No. 67 is unit No. 7 with a header, two-stretcher header vertical and horizontal border.

Bond Chart No. 68.

Bond Chart No. 68 consists of unit No. 8. It consists of crossed-stretcher courses on all odd numbered courses; and header, one stretcher, header, two-stretchers, on all even numbered courses.

Bond Chart No. 69.

Bond Chart No. 69 is unit No. 8 with a header horizontal and vertical border.

Bond Chart No. 70.

Bond Chart No. 70 is unit No. 9, consisting of header four-stretcher in every course.

Bond Chart No. 71.

Bond Chart No. 71 is unit No. 10. It consists of stretchers crossed on odd courses, and of header two stretchers on even courses.

Bond Chart No. 72.

Bond Chart No. 72. Flemish spiral bond. It consists of Flemish header courses, with the header laid over the joint in the course below.

Bond Chart No. 73.

Bond Chart No. 73. Zig-zag bond. It consists of header two stretchers in every course, the header being laid over the joint of the header in the course below.

GLOSSARY OF TERMS USED.

All-Stretcher Bond.—Bond showing only stretchers on the face of the wall, each stretcher divided evenly over the stretchers under it.

American Bond.—A name sometimes given to common bond.

Arches.—Bricks that are used to corbel out over the fires in the kiln.

Arches.—The brickwork over openings, or between skewbacks.

Ashlar Line.—The main line of the surface of a wall of the superstructure.

Backboard.—A temporary board on the outside of a scaffold.

Backing Up.—Any part or all of the entire wall except the overhand face tier.

Bat.—A broken piece of brick, generally a half a brick.

Batter.—A face of a wall leaning from the vertical toward the wall.

Batter Stick.—A tapering stick used in connection with a plumb rule for building battering surfaces.

Bed (of Brick).—The surface of the brick that should be laid downward.

Bed (of Mortar).—Mortar on a wall ready to receive a brick upon it. The mortar on which a brick rests.

Bed Joint.—The horizontal joint between two courses.

Benches.—Brick in that part of the kiln next to the fire, that are generally baked to vitrification.

Blocking.—A method of building two adjoining or intersecting walls not built at the same time, by which the walls are tied together by offset and overhanging blocks of several courses of brick.

Body Brick.—The best brick in the kiln. The brick that are baked hardest with the least distortion.

Bond.—The relative arrangement of vertical joints.

Boom Derrick.—A derrick having a boom supported by a mast, which in turn is supported by either stiff legs, guys, or both.

Brick-and-Brick.—A method of laying brick by which the brick are laid touching each other with only mortar enough to fill the irregularities of the surfaces.

Bull Header.—A brick laid on edge in the direction cross-wise of the wall. A header on edge.

Bull Stretcher.—A brick laid on edge in the direction lengthwise of the wall. A stretcher on edge.

Buttering.—Putting mortar on a brick with a trowel before the brick is laid.

Camber.—A convexity upon an upper surface.

Center.—A temporary support to masonry, such as to an arch.

Chimney Breast.—That part of the wall that is made to project from the line of the wall by a fireplace or by flues.

Closer.—A piece of brick laid to the line. Also, the last brick laid in any course of any tier.

Common Bond.—Several courses of stretchers followed by one course of either Flemish or full headers.

Corbel.—One or more courses projecting from the wall.

Core of Chimney.—The inside shell of the chimney.

Course.—One horizontal layer of brick in a wall. A radial layer of brick in an arch.

Crown.—Camber. The opposite of sag.

Cull.—One particular lot of brick sorted for color or size.

Culling.—Sorting brick for color, size or evenness.

Cutting-Out Hammer.—A hammer used for striking a chisel for cutting brick out of a wall.

Eastern Method.—The pick and dip.

Efflorescence.—A whitish deposit that sometimes appears on the surface of new walls.

Eight Inches (8″).—Two tiers.

English Bond.—Alternate courses of full headers and stretchers, the headers being plumb over each other, and the stretchers being plumb over each other. The headers are divided evenly over the stretchers, and over the joints between the stretchers.

English Cross Bond.—The same as English bond, except that the alternating stretcher courses, instead of being plumb over each other, break evenly over each other.

Face.—The front or exposed surface of a wall.

Face Brick.—Brick laid on the face of a wall.

Fat Mortar.—Mortar that tends to stick to the trowel, generally because of too little sand.

Filling In.—Laying brick on interior tiers after the face tiers in the same course have been laid.

Flat Arch.—An arch whose soffit is in approximately a level plane.

Flemish Bond.—Courses of alternate stretchers and headers, with the center of the headers located plumb over the center of the stretchers beneath them.

Flemish Cross Bond.—Any bond having alternate courses of Flemish headers and stretcher courses. The Flemish headers being plumb over each other and the alternate stretcher courses being crossed over each other.

Flemish Double Cross Bond.—Bond with odd numbered courses stretchers divided evenly over each other, and even numbered courses Flemish headers in various locations with reference to the plumb of each other.

Flemish Header.—One course of brick consisting of alternate stretchers and headers.

Footings.—The lowest courses of a wall.

Four Inches.—Tier. A vertical tier of brick the width of one brick in thickness.

Full Header.—A course consisting of all headers.

Garden Wall Bond.—A name given to any bond particularly adapted to walls two tiers thick. A bond consisting of one header to three stretchers in every course.

Green Brick Work.—Brickwork in which the mortar has not had time to set.

Grout.—A thin, soupy mixture of cement, sand and water.

Hand Leather.—A piece of leather used to protect the fingers.

Hard.—A term given to the brick that are thoroughly baked.

Header.—A brick laid so that only its end shows on the face of a wall.

Header Bond.—Bond showing only headers on the face, each header divided evenly on the header under it.

Header High.—The height up to the top of the course directly under a header course.

Jamb.—The end of a wall, as at a door or window opening.

Jointer.—A tool used for smoothing or indenting the surface of a mortar joint.

Key.—The center brick or course of brick of an arch.

Key.—The relative position of the headers of various courses with reference to a vertical line.

Lap.—The distance that one brick extends over the brick under it.

Laying to Bond.—Laying the brick of the entire course without a cut brick.

Laying Overhand.—Building the further face of the wall from a scaffold on the other side of the wall. Laying the outside face tier from the inside of the building.

Lead.—A part of the wall built up ahead of the line, to which to haul the line.

Lean Mortar.—Mortar that does not adhere to the trowel, generally due to the presence of too much sand.

Ledger.—A horizontal board nailed to the poles of a scaffold on which the pudlogs rest.

Light Hard.—A term applied to red brick that are not the hardest in the kiln. Although suitable for carrying moderate loads, they are not able to withstand alternate freezing and thawing as well as the hard brick.

Lime Putty.—Slaked lime without sand or cement.

Line.—The string used by the bricklayer as a guide for laying the top edge of brick.

Lintel.—A permanent horizontal support over an opening that may be curved or straight on the top.

Lipped.—Laid with a battering face.

Mercury Bob.—A plumb bob filled with quicksilver to get the greatest weight in the smallest size.

Mortar.—Any mixture used to fill the joints between bricks.

Mortar Bed.—A pen in which to make mortar.

Mortar Board.—A flat wooden board 3'0"x3'0" for holding mortar.

Motion Study.—The examination of the value, time and sequence of motions for producing the greatest results in the least time with the least effort and fatigue.

Offset.—A course that sets in from the course directly under it. Also called set-off, set-back, etc. The opposite of corbel.

Outrigger.—A joist projecting out of a window for supporting an outside scaffold.

Outside Four Inches (4").—The overhand face. The outside tier.

Overhand Work.—An entire wall built with a staging located on only one side of the wall.

Overhang.—A face of the wall leaning from the vertical away from the wall.

Pack.—A pocket and its load of two courses of bull headers.

Packet.—A tray for holding about 90 pounds of brick.

Peach Basket.—A templet against which the entire head of a tall chimney is built.

Peen.—That end of a hammer head which terminates in an edge.

Pick and Dip.—The name of the method where the bricklayer picks up a brick with one hand, and just mortar enough to lay it with a trowel with the other hand, simultaneously.

Pier.—An isolated masonry column, the brickwork between two adjoining openings in the same story.

Pilaster.—A pier projecting from a wall.

Pin.—An iron rod ⅞" by 10" to support the frame in the Gilbreth Scaffold horse.

Plumb.—Vertical.

Plumb Bob.—A line and weight for determining vertical lines.

Plumb Bond.—Another name for all stretcher bond work built with particular effort to have corresponding joints exactly plumb with each other.

Plumb Bond Pole.—A pole used for laying out the exact position of vertical joints.

Plumb Glass.—A slightly curved glass into which alcohol is sealed for use in a plumb rule.

Plumb Rule.—A tool used to aid in building surfaces in a vertical plane.

Pointing.—Pushing mortar into the joint after the brick is laid.

Pressed Brick.—Brick pressed in the mold by mechanical means before it is baked.

Pudlog.—A joist used for supporting scaffold planks. One end of a pudlog rests on the scaffold, and the other end rests on the face tier of a brick wall.

Quoin.—Brickwork in a corner.

Racking.—The method of building the end of a wall so that it can be built on and against without any toothers.

Rake.—The end of a wall that racks back.

Reveal.—The end of a wall, as at a jamb or return.

Rise.—The vertical distance between the level of the bottom of the skewback and the bottom of the key.

Rolled.—A brick laid with an overhanging face.

Routing.—Determining the way, the time and the method of getting materials from the point of shipment to the place where the workman puts them in place.

Rowlock.—Bull header—a ring of brick on edge forming an arch.

Run.—Planks used for workman to walk on.

Run of Kiln.—All brick in the kiln except those brick that are too soft or misshapen to be laid even in the filling tiers.

Running Bond.—Another name for all-stretcher bond.

Salamander.—A heater having no chimney.

Salmon.—Bricks that are softer than light-hard, and are suitable for little else than fire stopping.

Scale Box.—A derrick box made with an open top and one open end.

Segmental Arch.—An arch the bottom of which is the arc of a circle.

Set.—A wide bevelled edged chisel used for cutting brick.

Set In.—The amount that the lower edge of a brick on the face tier is back from the line of the top edge of the brick directly below it.

Set-off.—Set in.

Shanking.—Resting the hod on the end of the handle (or shank).

Shell of Chimney.—The outer wall of a chimney.

Shove Joints.—Vertical joints filled by shoving bricks as they are laid.

Sighting.—Observing with the eye the appearance of straightness of a line, such as the corner of a wall.

Skewback.—The line on the wall against which an arch is laid.

Slewing-rig.—The device used for swinging a boom derrick by machinery.

Slushed Joints.—Vertical joints filled by throwing in mortar with a trowel after the bricks are laid.

Soffit.—The under side of a covering over an opening, such as the bottom of a cap or arch over a window.

Spirit Glass.—The curved glass which contains alcohol or other non-freezing liquid in a plumb rule.

Spirit Plumb Rule.—A plumb rule with a curved glass nearly full of alcohol or other thin, non-freezing liquid. The location of the bubble with reference to a mark on the glass indicates the plumb position of the edge face of the plumb rule.

Spreader.—A temporary board put in a horizontal position half way up a window or door frame to prevent the masonry from crowding it inward.

Spring-stay.—A stay, made by two pieces of board separated by a brick, which holds a scaffold to a wall by the friction caused by the spring of the boards.

Stagings High.—About 3'8" with the Gilbreth Scaffold, about 5'0" high with the trestle horse.

Stock.—Brick and mortar.

Story Pole.—A pole on which all measurements of courses, openings, projections, off-sets, corbels, plates, and bottoms of beams of any one story are marked.

Straight Arch.—An arch whose soffit is in approximately a level plane.

Straight Edge.—A board having one or two straight and parallel edges, used for levelling and plumbing longer surfaces than can be reached with an ordinary spirit level.

Stretcher.—A brick laid so that only its long side shows on the face of the wall.

Stringing Mortar.—The name of a method where a bricklayer picks up mortar for a large number of brick and spreads it before laying the brick.

Struck Joint.—A joint that has the surface smoothed by a trowel.

Tapping.—Pounding a brick down into its bed of mortar with a trowel.

Template.—A pattern.

Tempering Mortar.—Softening mortar by adding water and stirring.

Tender.—A laborer who tends masons. A general name covering hod and pack carriers and wheelbarrow men.

Three-quarter Brick.—A brick clipped to about three-quarters its full length.

Tier.—A vertical layer of brick; four inches, or the width of one brick in thickness.

Toother.—A brick projecting from the end of a wall against which another wall will be built.

Toothing.—The temporary end of a wall built so that the end stretcher of every alternate course projects one-half its length.

Trestle Horse.—A four-legged horse.

Trig.—The brick midway between the leads that is used to support the line from sagging or vibrating due to many bricklayers constantly disturbing it.

Trimmer Arch.—An arch adjoining trimmer beams. The arch that supports a hearth to a fireplace.

Tub.—A half-barrel sometimes used in New England for holding mortar.

Twelve Inches (12").—Three tiers.

Two Inch Piece.—A closer about one-quarter of a brick in length used to start the bond from the corner.

Unit.—An arrangement of headers and stretchers which, repeated, forms definite bonds.

Up and Down.—The body brick together with the light hard brick.

Wall Ties.—Iron bands used to tie tiers of brick together or to tie the junction of two pieces of a wall, such as at corners, angles, and at toothing and backing.

Western Method.—The stringing mortar method.

Wire Cut Brick.—A brick having two of its surfaces formed by wires cutting the clay before it is baked.

Withe.—A 4″ partition or tie between two walls, such as two walls of a chimney.

INDEX.

Printed in Great Britain
by Amazon